Vernacular Regeneration

Urban regeneration is currently taking place in inner-city Johannesburg. This book presents an alternative, multi-layered account for reading the process of urban change and renewal.

The provision of social and affordable housing and the spread of private security are explored through the lenses of neoliberal urbanism, gentrification, the privatisation of public space and revanchist policing. This book interrogates these concepts and challenges their assumptions based on new qualitative and ethnographic evidence emerging out of Johannesburg. Dated concepts in Critical Urban Studies are re-evaluated and the book calls for an alternative, adaptable approach, focusing on how we develop a vocabulary and creative understanding of urban regeneration.

This book is an outstanding contribution to theoretical and comparative approaches to understanding cities and processes of urban change. It offers practical insights and experiences which will be of considerable use to practitioners, policy-makers and urban planning students.

Aidan Mosselson is currently a Newton International Fellow, based at the Department of Urban Studies and Planning, University of Sheffield. He completed his PhD in Social Geography at University College London in 2015 and was Associate Lecturer in the Sociology Department at the University of the Witwatersrand, South Africa. He held a Post-Doctoral Research Fellowship at the University of Johannesburg and the Gauteng City-Region Observatory. In 2017, he was awarded an International Fellowship by the Urban Studies Foundation, which supported a stint as a Visiting Fellow at LSE Cities.

Vernacular Regeneration

Low-income Housing, Private Policing and Urban Transformation in Inner-city Johannesburg

Aidan Mosselson

LONDON AND NEW YORK

First published 2019
by Routledge
2 Park Square, Milton Park, Abingdon, Oxon OX14 4RN

and by Routledge
52 Vanderbilt Avenue, New York, NY 10017, USA

First issued in paperback 2020

Routledge is an imprint of the Taylor & Francis Group, an informa business

British Library Cataloguing-in-Publication Data
A catalogue record for this book is available from the British Library

Library of Congress Cataloging-in-Publication Data
A catalog record has been requested for this book

ISBN 13: 978-0-367-58635-5 (pbk)
ISBN 13: 978-1-138-74694-7 (hbk)

Typeset in Times New Roman
by Integra Software Services Pvt. Ltd.

Contents

Illustrations

Figures

Maps

Table

Acknowledgements

Firstly, I would like to thank Ruth Anderson and Faye Leerink at Routledge for their assistance in bringing this book to light.

The majority of the research informing this book was conducted during the course of my PhD, which was generously funded by the Association of Commonwealth Universities and the Oppenheimer Memorial Trust.

I am indebted to all those who participated in my research. I have done my utmost to present the accounts, stories, feelings and points of view that you shared with me honestly and accurately. I am acutely aware that this work would not have been possible without the time and access people gave me. I have endeavoured to produce work which can make a contribution, be it through informing policy and governance agendas, contributing to public debate or simply sharing and honouring certain people's stories, and thus repay some of the kindness and generosity which my research participants showed towards me.

Most of the work that went into the book was done whilst I was a Post-Doctoral Research Fellow (PDRF), hosted jointly between the University of Johannesburg and the Gauteng City-Region Observatory – a partnership between the University of Johannesburg, the University of the Witwatersrand and Gauteng Provincial Government. I was hugely privileged to be given the time, space and freedom to work on my own research and writing agenda, and had I not received the PDRF position this book would never have been possible. This sort of opportunity is all too rare for early-career academics, and should be made available on a far wider scale. I would like to thank Rob Moore and Graeme Gotz for giving me space to pursue my work and trusting me to manage myself and my workload. I am also grateful to Richard Ballard for his mentorship and guidance, Sian Butcher for her company and intellectual insights, Adele Underhay, Farah-Naaz Moosa, Nadine Abrahams, Elaine Milton and Ruth Mohamed for making sure everything ran smoothly and we were always well fed, and the very talented Mncedisi Siteleki for designing the maps which feature in the book. Photographs were kindly provided by Thembani Mkhize, who also played an important role in some of the later research, and whose work on Hillbrow and building managers provided valuable insights and points of learning for me. I would also like to express my appreciation to all the other members of staff at the GCRO (especially the members of the Innovation Hub!); my time there was enjoyable and I value the friendships which we made.

I was extremely fortunate to be granted an International Fellowship by the Urban Studies Foundation, which facilitated a visit to LSE Cities in the latter stages of writing. I am extremely grateful for the financial support provided by the Urban Studies Foundation and would like to thank them for the faith they placed in me and this book.

My time at LSE Cities was brief, but extremely productive. I was most fortunate to work under the guidance of Suzanne Hall, whose work, personal kindness and political convictions have long been inspirations for me. I would also like to thank Ricky Burdett and Philipp Rode for supporting my visit, Samuel Scott and Andrew Sherwood for all their support and assistance, and the team at LSE Cities for making me feel welcome.

My intellectual trajectory has been shaped by many people, but I am particularly grateful to Suzi Hall (again), Jennifer Robinson and Charlotte Lemanski who have provided me with ongoing support, mentorship and encouragement. I hope that I have merited the time and effort they have committed. I would also like to thank Claire Colomb, Claire Bénit-Gbaffou and Richard Ballard, who all played important roles in reading my work at various stages and helping me refine my thinking around particular issues.

Finally, I would like to thank my family for their constant love, support and encouragement. I would also like to thank my friends in Johannesburg and London. Above all, I have to thank Lioba Hirsch for being my source of motivation, intellectual companion, best friend and love.

An earlier version of chapters 5 and 6 appeared in '"It's not a place I like, but I can live with it": ambiguous experiences of living in state-subsidised rental housing in inner-city Johannesburg,' *Transformation: Critical Perspectives on Southern Africa*, 2017, 93, 142–169. Some of the information regarding urban regeneration and housing policy in chapters 1 and 2 can be found in my chapter 'Caught between the market and transformation: urban regeneration and the provision of low-income housing in inner-city Johannesburg,' in P. Watt and P. Smets (Eds.), *Social Housing and Urban Renewal: A Cross-National Perspective* (Emerald, 2017). Earlier parts of Chapter 7 were published in '"Joburg has its own momentum": towards a vernacular theorisation of urban change,' *Urban Studies*, 2017, 54 (5), 1280–1296.

Acronyms

AFHCO	Affordable Housing Company
BBP	Better Buildings Programme
BID	Business Improvement District
BNG	Breaking New Ground
CBD	Central Business District
CDS	City Development Strategy
CID	City Improvement District
CJP	Central Johannesburg Partnership
CPF	Community Policing Forum
DA	Democratic Alliance
GEAR	Growth Employment and Redistribution
GPF	Gauteng Partnership Fund
ICHUT	Inner City Housing Upgrading Trust
ICRC	Inner City Regeneration Charter
ICRM	Inner City Road Map
JDA	Johannesburg Development Agency
JHC	Johannesburg Housing Company
JOSHCO	Johannesburg Social Housing Company
JSDF	Johannesburg Spatial Development Framework
NASHO	National Association of Social Housing Organisations
NGO	Non-Governmental Organisation
NHFC	National Housing Finance Corporation
RCID	Residential City Improvement District
RDP	Reconstruction and Development Programme
TUHF	Trust for Urban Housing Finance

1 Thinking with and through Johannesburg

Introduction

This is a book about urban change. Cities are dynamic, ever-changing places. As such, they require agile, flexible theoretical lenses and analytic repertoires (Robinson, 2016). But the change which occurs in each specific city is not open-ended or indeterminate. Cities change within the bounds established by their histories, the contemporary socio-political juncture at which research is taking place, and within enduring topographies and physical spaces. This book, then, is about understanding the ways in which processes of urban change unfold in a specific place, with a specific set of histories and dynamics. It pays close attention to the urban regeneration project currently underway in Johannesburg's inner-city, and seeks to understand the range of factors and actors which have come together to shape the process and the effects that it is having on the lived realities unfolding in the area, 24 years since the end of apartheid. The research and discussion cover several areas and topics, focusing on policy formation and the relationship between state intervention, ideologies of redistribution and market-based approaches to urban redevelopment; a sociological study of various housing developers and investors; an examination of everyday policing and security provision in a fraught, stressed urban environment; and a reflection on the lives and experiences of lower-income residents who make the area home. The book attempts to tie these different strands together to develop a fuller picture of experiences of urban change, and to understand what they mean in the context of the post-apartheid city.

Although the book is focused on a specific case, the world of cities is not fragmented, and there is increasing recognition that cities everywhere are linked through relational networks, connections and flows of people, capital and ideas. In examining one particular process of change, the book also asks what can be learnt from Johannesburg that can help us understand dynamics in other places too. It therefore explores the relationship between specificity and generality, and the extent to which it is possible to develop theory which can travel from one individual (although certainly not bounded or isolated) case to speak to and aid in making sense of other settings and localities. It is thus an attempt to apply, as well as explore the limits of, a comparative imagination.

Speaking to elsewhere from Johannesburg is certainly not an obvious or straightforward undertaking. Although iconic, Johannesburg is, on the surface, far from the norm. It is a city which is most frequently noted for its archetypal extremes – its alarming inequalities, distorted and fragmented urban form, severe levels of violence and stark, enduring racial divide. These are all certainly specific problems which affect life in the city and continue to harm many of its approximately five million residents. However, these are also problems which are not unique to Johannesburg. Fear of violence and insecurity, racial discrimination, rising inequalities and housing shortages are shared experiences for urban dwellers across the globe. Thus, insights gained in Johannesburg do not only speak to the immediate context, but can be helpful for thinking through or analysing other places too. At the same time, Johannesburg is an ordinary city like any other, where people, both despite of but also because of the above-mentioned disparities and forms of violence, go about their everyday lives, seeking ways to feel safe, to raise their families, to find work and shelter, to build homes, to connect with others and to get on with not only surviving, but also making the most of life. Some people also engage in the extremely difficult work of making the post-apartheid city better, of attempting to develop housing, enact planning procedures which mitigate fragmentation and exclusion, create employment, govern the city for all its residents, care for friends and family, and even heal racial divides. It can be instructive, then, to look more closely at how people undertake these endeavours in an 'extreme' context, to both reveal some of the difficulties which characterise life in other places too, but also to learn about the creativity, urgency and innovation which go into urban governance, and that make up and sustain everyday lives.

The not-so-extreme city

Whilst undertaking the research which informs this book, I found my way into a range of different places. Buildings, parks, recreation facilities, churches (some of which used to be synagogues), streets crowded with people, the back of a police van (also crowded with people), shopping malls, a gym, corporate offices and homes. Some of these places certainly exemplified the 'extremes' which Johannesburg is noted for. For example, Pullinger Kop Park, located on the eastern edge of Hillbrow, is a desolate patch of land and widely known 'hotspot' where drugs are easily available. Built on a plot that was once the stately home of a man who made his fortune in the brutally exploitative mining industry, and still bearing his name, it underscores how the country's system of racialised capitalism generated luxury and vast sums of wealth for some people, whilst simultaneously impoverishing the black[1] majority and leaving ruination in its wake (cf. Stoler, 2008). As the 20th century unfolded, the wealthy elite moved away from the central city and began to settle in the city's northern expanses. The land on which Pullinger's house rested was donated to the City and turned into a public park. The movement of wealthy white people away from the central city prefigured the area's decline and the destruction of most public infrastructure in the vicinity. The fate which befell the park thus also clearly shows how capitalism uses people and places, and then discards them once value can no longer be extracted from them (Chari, 2017).

Today, the park is in a severely dilapidated state. When I visited, it was a cold Saturday night in the middle of the bitter Highveld winter. Despite the cold, the park was crowded with people, the majority of them homeless, buying and using the available drugs, and then retreating to the cardboard boxes and assorted blankets which served as their makeshift homes. The scene in the park underscored the severe, seemingly intractable problems which continue to define many people's lives in Hillbrow. It made vivid the ways in which chronic poverty, unemployment and homelessness are the reality for far too many people, despite the city's general wealth and prosperity. It also clearly demonstrated how people, lacking any other forms of social and physical support, depend on cheap, harmful substances to make it through the cold night, as well as how others, lacking other employment opportunities, prey on their vulnerability to earn income. The events unfolding in the park also brought home the ostensible impossibility of governing and improving the inner-city. The bulk of the drug trade was taking place in a basketball court which had been installed in the park during a previous attempt to improve the neighbourhood; the only remaining traces of the improvement effort were a sign announcing the 'Pullinger Kop Revitalisation Project,' which had long-since been effaced and scribbled over, the rims of the baskets at each end of the court – the supporting structures having been thoroughly stripped of any metal or other useable material – and the paved surface that people were bedding down on for the night. The despondency of the situation was overwhelming, and I found myself at the limits of my theoretical vocabulary and ability to make analytic sense of it all. At that point, my eyes fell upon a piece of writing scrawled in a shaky hand onto the 'renewed' surface of the basketball court. In large black letters, someone had written the word 'FUCK,' which, I had to admit, was the most eloquent way of summing up the situation.

Although Pullinger Kop was a particularly difficult, disturbing setting and vivid reminder of Johannesburg's discarded people and spaces, it was also just one of many different spaces which make up the inner-city. Another park, also located in Hillbrow, captures some of the significant changes which have taken place, but also the mundane, but no less significant, effects they have had. Ekhaya Park, which features prominently in latter chapters, is a place I grew familiar with and often look back on fondly. Jointly funded by the Ekhaya Residential City Improvement District, the Johannesburg Development Agency and FIFA's World Cup legacy programme, it is an AstroTurf football pitch surrounded by a tall wire fence and accompanied by a playground and sets of benches. It was built on a plot of land which, like many other spaces in the neighbourhood, had fallen into disuse. I attended several Ekhaya Kidz Soccer Days at the park, where teams comprised of children from surrounding residential buildings compete in day-long tournaments, cheered on by their carers and neighbours. These days were always heart-warming occasions, and offered powerful juxtapositions to the desolation of the other park. Even on days when there weren't organised events taking place, the football pitch was always in use, with children playing, shouting and just enjoying being kids, in what was once one of the most dangerous, violent neighbourhoods in the city (Leggett, 2003).

Whilst not taking away from the exclusion, suffering and hardships which characterise life for far too many of Johannesburg's residents – particularly those who shelter in the city's derelict buildings, its parks (in the inner-city and suburbs) and in the informal settlements which dot the city's periphery, but have also sprung up in some abandoned warehouses and industrial buildings on the edge of the inner-city – it is the ordinary, not-so-extreme spaces, such as Ekhaya Park, that I want to draw attention to. When writers describe Johannesburg as a 'city of extremes' (Murray, 2011), they draw our attention to the continuing racial disparities and inequities which define contemporary South Africa, and the ways these are expressed through the divide between spaces of (predominantly white) privilege, such as pristine, exclusive shopping malls, fortified, luxurious gated communities, insulated office parks and corporate sky-scrapers, and homelessness, informal settlements, chronic, lifelong unemployment and relative deprivation, particularly in the former black townships. Whilst disturbing, these stark visions fail to include the in-between, not-so-extreme spaces that also make up the city. Whilst a lot of my research took place in public spaces, including the two parks discussed above, a great deal was also carried out inside residential buildings. These interiors offer important insights into the combination of somewhat drastic measures and approaches to security which define the regeneration process, as well as the ordinary, everyday lives which are unfolding inside them.

Places which stand out in my mind, several years after the bulk of the research was conducted, include a freshly painted, open-plan room, with a small built-in sink and hotplate that was the soon-to-be home of two adult men, both of whom work as security guards and need a way to stretch their meagre salaries; a largely empty, two-bedroom apartment in Hillbrow whose long-faded parquet floors recalled the previous glamour and desirability of the neighbourhood, where, for want of other furniture, my host and I sat on upturned beer crates as we discussed life in the area; a one-bedroom flat where the living room doubles as a stay-at-home father's office and his young child's bedroom; another one-bedroom flat where a large cabinet, complete with a flat-screen TV and speaker system, serves as a divider between the living room and children's bedroom. These spaces stand out for several reasons: firstly, they demonstrate the flexibility which is required to make homes in the inner-city – sub-dividing apartments, sharing a room with a friend, relative or colleague, sacrificing privacy in order to afford central, well-located accommodation – as well as the work that goes into making a life – filling a room with objects such as televisions, speakers, couches, religious artefacts, but, more importantly, carving out a space of quiet and safety where children can be raised and weary bodies can be rested, away from the constant noise, bustle and movement of the throbbing streets outside.

Secondly, paying attention to the everyday ways in which people construct homes for themselves draws attention back to the agency which people possess in all areas and facets of life, and which often gets neglected in extreme accounts of relentless poverty and suffering (Back, 2015). Accounting for these forms of agency is not intended to detract from broader structural problems, including institutionalised racism and inequality, enduring patterns of colonial violence and dispossession, and the damage wrought by neoliberal governance agendas. It is, however, intended as a salve to the more despairing images of life in the post-apartheid period, where constant emphasis on policy failures, inefficiency and corruption, governmental neglect and profit-driven ideologies frequently obscure the ways in which millions of people's lives have changed for the better. These narratives also discredit the ingenuity and experimentalism which are part and parcel of urban life, and the shifting roles and identities which people adopt as they try to make meaningful improvements to the city. Lastly, attending to the not-so-extreme spaces of the city helps situate the experiences of black urban dwellers in the story of ordinary cities around the world (Robinson, 2006), and demonstrates that the lives which people construct in African urban contexts are not only shaped by the extremes of poverty, deprivation and injustice (although these are central parts of the story), but are also based on mutual support, care and simple efforts to get by. Because they are less extreme than we are often led to believe, the lives people lead and choices they make are also marked by practices and lessons which people concerned with other contexts can draw on, relate to, and learn from.

Situating Johannesburg's inner-city

Socio-political landscape

The interiors I am describing above are important because of their physical and socio-political location. Johannesburg's inner-city has always exemplified and materialised the trends which have shaped the country at its various stages of existence. The modern city, seen from above in Figure 1 below, grew out of the gold rush at the end of the 20th century, and bore the hallmarks of this lucrative, but also highly extractive, violent and exploitative industry. The richest seam of gold ever discovered gave birth to a city at the forefront of modernity, as skyscrapers quickly begun to sprout on a once dry, dusty patch of land (Beavon, 2004; van Onselen, 2001). Street names such as Prospect, Nugget, Claim and Quartz still pay homage to the mining industry and mark the area's geography and people's routes through it (Nuttall, 2004). Less visible, however, are the places of harassment, imprisonment and violence which constituted the black population's experiences of the colonial and apartheid city. Although memorialised in places such as Constitutional Hill and the Drill Hall, the every-day violence and oppression which marked the city and through which its 'European'[2] character was kept intact are not always readily apparent. However, the scrutiny, policing and exclusion which black people were subjected to did as

Figure 1.1 Aerial image of inner-city. Photograph by Clive Hassel.

much to define the city as its burgeoning wealth and glamour (Mbembe, 2008). In that sense, Johannesburg has always intertwined two elements: modernism, aspiration, commerce and wealth, on the one hand, and brutality, repression, white racism, exploitation and subjugation, on the other.

Riding on the high price of gold, the city flourished through the middle decades of the 20th century. The wealth flowing from the mines became concretised in the city's built form, with older buildings being demolished and the bulk of the high-rise buildings and skyscrapers which make up the city's contemporary skyline dating from this period, through to the 1970s (Chipkin, 1993). However, as the apartheid regime's grip on power became increasingly unstable, the fortunes of the inner-city began to decline too. Downturns in the price of gold, international sanctions, deindustrialisation without concomitant reinvestment, and political insurrection all put pressure on the apartheid government and began to change the fabric of the central city. As investors grew wary about the political situation, they began to move away from the inner-city, with factories, warehouses and shops closing down, corporate headquarters beginning to relocate and wealthier white residents leaving for the leafy suburbs (Beavon, 2004; Goga, 2003; Morris, 1999a). Whilst this process of flight had deleterious consequences for the built environment, with neglect and ruin becoming increasingly prevalent, it also signalled a broader change. Before apartheid officially came to an end, more and more black people began to move into the inner-city.

What was once the bastion of white, European urbanity started to more closely reflect the African context in which the city is situated. Newly arriving residents marked the inner-city with their presence, and a new sense that the city was moving away from, but still very much influenced by, its European, racist origins took hold (Mbembe and Nuttall, 2008; Tomlinson et al., 2003).

When apartheid formally came to an end in 1994, Johannesburg's inner-city was markedly different, demographically, but also in terms of how the built environment was used and lived in. In a short space of time, wealthy white people were replaced by black residents, who were generally poorer, and many could only afford the rents being charged by sub-letting and sharing rooms with far more people than they were originally built to house (Crankshaw and White, 1995; Morris, 1997). Landlords who remained in the area (and didn't simply abandon their properties in the face of drastic capital flight and residential change, as many did) exploited their tenants and the growing demand for accommodation, charging high rentals and not investing in any maintenance or upkeep (Morris, 1999b). The area's economy also changed, with corporate and financial services giving way to small-scale manufacturing and survivalist trading. These changes infused the area with a different character, but also brought enhanced governance challenges, forms of poverty and deprivation, struggles over space and rights to belong, and crime and violence. At the same time, the changes which were affected also recreated formerly segregated places like Hillbrow, Berea and Yeoville as sites of emancipation, mobility, adaptation and experimentation (Gotz and Simone, 2003). As these changes are lived out in the shadows of the former colonial and apartheid urban landscape, Johannesburg has come to embody the hopes, optimism and dynamism, but also the challenges, legacies, inequities and forms of deprivation which define South African urban modernity. Today the city stands as a representative of both South Africa's apartheid past, as well as its uncertain, still-emerging postcolonial present.

Perhaps more than any other setting in Johannesburg, the inner-city demonstrates the changes which have occurred since the end of apartheid. Today, it is a mix of old and new infrastructures and is animated by a diverse population. In a generally dispersed urban landscape, it is densely populated, accounting for roughly 12% of the City of Johannesburg's population (of almost five million) (HSRC, 2014). Although the area has been largely abandoned by corporate South Africa, it still has an immense economy, with vast amounts of commodities and cash circulating through it on a daily basis (Zack, 2016). Whilst South Africa's major banks maintain offices in the area, they have all relocated their corporate headquarters to other, more illustrious parts of the city. Anglo-Gold Ashanti, one of the largest gold producers in the world, have remained in the area, basing themselves in Marshaltown, a banking district which has emerged on the site of the original town centre, and where an open-air museum commemorates the city's inextricable link with gold mining. Outside of this small precinct, small-scale manufacturing, motor mechanics and trading, both formal and informal, are the predominant forms of economic activity.

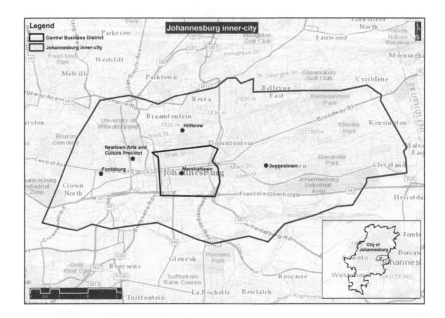

Map 1.1 Map of inner-city.

The inner-city's social geography is diverse. Various populations have imprinted their presence on distinct areas: Fordsburg, on the western side of the inner-city, is a semi-industrial, low-density suburb traditionally home to South Africans of Indian origin. It boasts a number of restaurants and cafes purveying Indian cuisine, and Bollywood movies and assorted paraphernalia are widely available. More recently, migrants from Pakistan, Bangladesh and Somalia have settled there, drawn to its mosques and Islamic community organisations (Dinath et al., 2014; Jinnah and Rugunanan, 2016). Running eastwards, Braamfontein is a commercial district with a large supply of high-rise office space. It is also increasingly providing accommodation for students, as the University of the Witwatersrand is located within this district and the University of Johannesburg is also nearby. The burgeoning student population has attracted new investments, and several popular bars, entertainment venues, boutique shops and markets have opened in recent years. This area has captured local and international attention, and tourists, aspirant photographers, filmmakers and artists have become ubiquitous. However, whilst the area profits from the presence of the universities and their students, fractious political movements and events originating in the universities have affected the broader vicinity too. In 2016, demonstrations over the costs of higher education at the University of the Witwatersrand spilled out into the streets surrounding the university, and the police fought running battles with students throughout

Braamfontein. The police fired numerous rounds of teargas and rubber bullets, several buses were set ablaze and some shops were looted.[3] These events were reminders that frustrations with ongoing inequality, blocked social mobility and persistent institutionalised racism cannot be confined to the universities, and are increasingly part of South Africa's urban fabirc. In this case, the city was not just a setting, but a vital staging ground through which these politics could be expressed.

The politics of the moment also play out in the spaces of the Central Business District (CBD). This was originally the centre of the economy during the gold mining era, home to numerous corporate headquarters, financial firms and mining companies. Today it is a mixed-use area, comprising commercial and retail services, both formal and informal, as well as residential buildings and the city legislature. As white South African-owned businesses have vacated the area, different groups of African migrants have come to capitalise on the commercial opportunities available. Jeppe Street, for instance, has attracted a host of Ethiopian businesses and traders. Known colloquially today as 'Little Addis,' entire buildings have become home to shops and warehouses trading in goods from East Africa, including coffee, spices, Ethiopian music, cooking utensils and clothes. In these spaces, the ways the city has been reabsorbed into the African continent become palpable. The effects of migration from the rest of the continent also make themselves felt in and around the Central Methodist Church. Neighbouring the Johannesburg High Court, the church has become a place of refuge for migrants, predominantly from Zimbabwe. In 2008, people fleeing outbreaks of violence targeting African migrants found their way to the church, which opened its doors and became a makeshift refugee centre. Xenophobic hostility towards other Africans has not yet subsided, and the church continues to shelter countless people, much to the chagrin of local government and the neighbouring legal professionals, businesses and property owners (Kuljian, 2014).

Demographic, economic and social change are also palpable in the high-rise suburbs of Hillbrow and Berea, which adjoin the CBD on its north-eastern edge. Occupying an area roughly 2 km^2, these areas have some of the highest residential densities in the entire city and are presently home to approximately 200 000 people. Whilst they were originally established to house the white population employed in and around the CBD, today they are home to a predominantly black African population, comprising South Africans as well as migrants from other African countries, mostly Zimbabwe and Nigeria. Hillbrow is a particularly notorious area and is known throughout South Africa, by black and white people alike, for its high levels of crime, violence, sex work and drugs. North of Hillbrow are the lower-density residential suburbs of Yeoville and Bellevue, which are also home to large migrant populations, predominantly from the Democratic Republic of Congo (Prabhala, 2008). The northern edge of Yeoville is framed by Louis Botha Avenue, one of the main thoroughfares in Johannesburg, which runs all the way from the inner-city to Alexandra township in the north-east of the city. It is one of the busiest roads in the country and has become part of the ambitious Transit Oriented Development strategy currently being implemented in various sites across the city (see Rubin and Appelbaum, 2016).

The eastern side of the inner-city is largely made up of warehouses and factories. Ellis Park Stadium, the site of South Africa's famous victory in the 1995 Rugby World Cup, and the Johannesburg Athletics Stadium are also located in the industrial suburb Doornfontein. Jeppestown is a suburb in the eastern part of the inner-city and is proliferated with disused industrial buildings. The city's disturbing poverty makes itself felt here, as many of these buildings have become people's homes, despite lacking ventilation, water and electricity (Mayson and Charlton, 2015). At the same time, the section of Jeppestown closest to the CBD is home to the Maboneng Precinct, an area in which former industrial buildings have been converted into artists' and fashion designers' studios, high-end retail outlets, restaurants and residential units. It currently occupies 150 000 m^2 and houses roughly 500 residents, but the developers have ambitious expansion plans (Propertuity, 2013). The area south of the CBD and Jeppestown is known as City Deep. It bears the strongest traces of the city's mining industry and is proliferated by disused mining land and the city's iconic mine dumps which, along with the freeway, also form the southern border of the inner-city. Despite highly toxic soil and water, communities of worshippers have claimed spaces around the abandoned mines, establishing churches and holy sites for themselves, thus reanimating spaces of exploitation, abandonment and ruin and investing them with new hopes, dreams and aspirations (Malcomes and Wilhelm-Solomon, 2016).

Decline and revitalisation

All of the changes which have taken place in the inner-city were made possible by a preceding, drastic process of capital flight and decline. Beginning in the 1980s, white-owned businesses began to relocate away from the inner-city. As international sanctions began to bite and internal political activism made apartheid increasingly untenable, many investors started to feel uncertain about the country's political and economic stability. Simultaneously, increasing numbers of black people began to move into inner-city neighbourhoods which were deemed off-limits to them. They were compelled to move away from the racially designated townships where housing shortages were reaching critical dimensions (Crankshaw and White, 1995). Furthermore, following the 1976 student uprisings in Soweto, the political climate in the townships had become increasingly fractious and these areas became intense sites of conflict between residents and the apartheid security forces. They became increasingly volatile and riven by police brutality and violence between different political factions, prompting many of those who could afford to live elsewhere to leave.

Initially Indian and coloured people began to move into inner-city suburbs such as Hillbrow. The majority who settled in Hillbrow in this period were middle class, and were generally employed in clerical and administrative roles (Crankshaw and White, 1995). Over time, they were followed by greater numbers of Black Africans. Due to apartheid's systematic differentiation according to racial categories, Black Africans were (and continue to be) the most impoverished group in South Africa. Those who moved into Hillbrow at this stage were therefore less educated and earned lower salaries than the Indians and coloureds who arrived before (ibid.). Faced with increasing changes in the area, white residents began to leave in great numbers. Not only were they motivated by fears couched in racism, they were also enticed by the government's policy of providing financial incentives for aspiring white professionals to move into the decentralised suburbs, particularly in the northern regions of the city (ibid.).

In growing numbers, black people in dire need of accommodation rushed in to fill the openings created by fleeing white residents, despite the threats of removal they faced from Influx Control and Group Areas laws.[4] Many landlords were quick to seize on the opportunities presented by the new demand for inner-city accommodation, some capitalising on the precarious situation of their newly arrived tenants, exploiting them by overcrowding apartments and refusing to maintain their properties (Morris, 1997). A spiral of neglect and destruction ensued, leading to the development of slum-like conditions in many inner-city buildings. Financial institutions exacerbated the situation as, growing increasingly jittery about the fate of the inner-city, they began red-lining the area, refusing to provide finance for property investments. As a result, owners wishing to maintain their buildings frequently could not raise the capital to do so, and higher-earning people were disincentivised from remaining in or moving into the area. Influx control laws were eventually repealed in 1986, greatly accelerating the movement of black people, particularly Black Africans, into the inner-city and causing the demographic composition and character of the area to change fundamentally (Morris, 1999c).

Whereas in 1986 only roughly 20 000 of Hillbrow's 120 000 residents were black, by 1993 85% of the residential population was black and by 1996 only 5% was white (Tomlinson et al., 2003). This was a pattern repeated in the central regions of cities throughout South Africa, as apartheid began to be dismantled and new urban patterns started to arise (see Maharaj and Mpungose, 1994). Faced with declining infrastructure, the increased presence of populations previously regarded as 'undesirable' and un-urban (Robinson, 1996) and rapidly spreading crime and grime, the flight of businesses and capital from Johannesburg's inner-city accelerated, culminating in 1998 in the relocation of the Johannesburg Stock Exchange to Sandton, a rapidly expanding commercial suburb in the north-eastern part of the city. This was also a period in which property speculation in the northern regions of the city was growing and investors were increasingly being encouraged by local government to look to suburbs like Sandton for new opportunities and profits (Goga, 2003). The decision to move the Stock Exchange confirmed the ascent of the northern districts of the city as the new financial centre and effectively delivered the 'knockout blow' to the old CBD (Beavon, 2004, p. 259), leaving it to fall further into disrepute and decay.

There have been several efforts to arrest the decline, some more successful than others. Early efforts aimed at turning the inner-city into both a trading hub, imaging it as the 'gateway to Africa,' and a high-tech financial centre (Bremner, 2000). These ambitious plans have given way to more pragmatic, but, as will be argued throughout this book, more durable and substantial renewal plans. The revitalisation efforts which have had enduring effects on the inner-city, and which this book is concerned with, have focused on the provision of social and affordable housing, largely through refurbishing or converting disused and derelict residential or commercial buildings. The social housing sector is dominated by private, non-profit companies, although there are several state-owned social housing institutions active around the country, including the Johannesburg Social Housing Company (JOSHCO), who are based primarily in and around the inner-city. Social housing caters to households whose salaries are above the maximum requirement to qualify for free state-provided housing, but remain unable to access accommodation through commercial markets. On current figures, those qualifying to be housed by social housing institutions earn between R3500 and R7000 per month. Domestic workers, security guards, waiters, shop assistants, low-skilled manual labourers or self-employed informal traders or craftsmen are frequently tenants in social housing developments.

Affordable housing is a more controversial category. Private companies branding themselves as 'affordable' make up the bulk of the formal rental market in the inner-city. Their rentals are targeted at households earning between R3500 and R14000 per month, and generally appeal to people employed in stable but low-paying jobs, such as teachers, police officers, professional assistants, nurses, call-centre workers and bank tellers. However, the 'affordability' of these offerings is questionable. Income levels vary within the inner-city: there is a sizeable low-to-moderate income population, as 21% of inner-city households earn monthly salaries between R3500 and R7500; there is also a growing segment that can be thought of as middle class, with 17% earning between R6366 and R12816 per month. However, 49% of inner-city households earn less than R3200 per month and there is a 25% unemployment rate (SERI, 2013). Thus, whilst a significant proportion can afford the rents being charged, private-sector accommodation is beyond the reach of the bulk of population residing in the area. The discrepancy between the rents charged and what people can actually afford to pay has led inner-city residents to find alternative, improvised living arrangements, including under bridges, in derelict buildings lacking water and electricity connections, and in former industrial buildings which have been haphazardly converted into living spaces (Mayson and Charlton, 2015).

At the same time, as I will demonstrate, housing providers too have had to adjust to the inner-city's realities and find inventive strategies for navigating between commercial demands and the constraints faced by the bulk of the people who call the area home. These acts of navigation, pragmatism and adaptation are at the core of my concerns in writing this book, as I seek to explore how pursuing urban regeneration in the context of Johannesburg's inner-city (and other locations too), requires flexibility, cognisance of the prevailing socio-spatial circumstances and inventiveness. Focusing on these activities helps make sense of the process underway, and account for, as well as critique, some of the effects that it is having on the area and the people living in it. As will hopefully become clear, the book aims at providing an inductive account of the regeneration process, drawing on the multiple motivating logics and agendas which have inspired it, the practical activities which go into grounding improvements in a volatile environment, and documenting the ways people live through and with processes of urban change. The overall goal is to emphasise the need for research to engage with and reflect local concerns, agendas and life-worlds, and to avoid over-reliance on pre-given categories or analyses. In turning to Johannesburg, I call attention to the ambiguity and diversity of processes of urban change and emphasise the unpredictability of vernacular experiences and practices everywhere.

Contemporary urban governance: between neoliberalism and developmentalism

By following the regeneration process underway in Johannesburg, this book aims to contribute to the burgeoning debates about urban developments in contemporary South Africa. South African cities have provoked a range of considerations and interventions. Studies have focused on numerous issues, not all unique to South Africa, but often presenting themselves in intense forms: increasing inequality and urban segmentation (particularly through gating and neighbourhood enclosures) (for example Clarno, 2013; de Vries, 2008; Dirsuweit and Wafer, 2006; Harrison et al., 2003; Murray, 2011); governing, or in some cases seeking to eradicate, informality (for example Lemanski, 2009; Mitchell and Heynen, 2009; Steck et al., 2013; Tissington, 2009); housing provision and policy shortcomings (for example Charlton, 2009; Charlton and Kihato, 2006; Huchzermeyer, 2001, 2014; Lemanski, 2006; Oldfield and Greyling, 2015); governance, policy and infrastructural innovations (for example Harrison et al., 2014; Harrison and Harrison, 2014; Rubin and Appelbaum, 2016; Todes, 2014); as well as increasing numbers of studies which engage with changing urban contexts through cultural and literary lenses (for example Charlton, 2017; Hlongwane, 2006; Matsipa, 2017). In terms of the urban policy literature, which this book draws on and aims to contribute to, there are two broad categories or schools of thought. The first focuses on the advent and effects of neoliberal interventions in the post-apartheid period. The second has arisen as a response to the former, and has come to stress the ongoing experimentation and developmental slant of governance in the contemporary period.

Neoliberal urban governance after apartheid

In the aftermath of its decay, Johannesburg's inner-city became the object of ambitious renewal plans. Broadly, these state-led initiatives sought to rebrand the area as the 'Golden Heartbeat of Africa' and a 'World Class African City' (Bremner, 2000). They thus sought to create an entrepreneurial environment which would attract local and overseas investment and are captured by the aspiration to World City status (Lipietz, 2008; Tomlinson et al., 2003). For many critics, these ambitions and the restructuring of local government which accompanied them clearly signal the advent of neoliberal forms of governance in the city. As Sihlongonyane (2015, p. 1620) concludes, the initial regeneration strategies developed for Johannesburg clearly signal the dominance of a pro-growth agenda and are 'inimical to the annihilation of economic inequalities.'

For instance, whilst the City Council's short-to-medium term planning and strategy document, *iGoli 2010*, emphasises the need for redistribution, equitable delivery of services and resources, and the enhancement of deprived people's access to economic opportunities, overall, the focus remains augmenting Johannesburg's competitiveness and attractiveness as an investment destination (Johannesburg Development Agency, 2010). Similarly, The Inner City Charter, the document which provides the basis for the renewal strategies undertaken in the 2000s, firmly commits to a pro-growth agenda, and frames the underlying ambitions behind regeneration as increasing business opportunities and property values in the area (City of Johannesburg, 2007). In these ways, the strategies draw on and continue the path set out in the Growth, Employment and Redistribution (GEAR) macroeconomic programme. Adopted in 1996, GEAR emphasises the need for government to maintain fiscal discipline and focus on the cost-effectiveness of its activities (Bond, 2000). It still maintains a commitment to social redress and the redistribution of economic wealth to communities excluded and impoverished under apartheid, but focuses on achieving these aims through 'trickle down economics,' where economic growth and competiveness are prioritised as the preconditions required to achieve wider social goals (Gumede, 2007).[5] GEAR is also the framework which established privatisation as an official component of economic policy, focusing on the need to sell-off state assets, establish public–private partnerships and run state activities as profitable parastatals (ibid.).[6]

There are direct continuities between the principles of GEAR and those adopted by the City of Johannesburg in the late 1990s and early 2000s. For instance, following the restructuring process recommended by the strategic document *iGoli 2002*, the provision of municipal services has been taken over by utilities which are registered as separate companies, such as Johannesburg City Power, Johannesburg Water and Pikitup (which provides refuse removal, recycling and waste disposal services). These public–private companies operate on cost-recovery principles, and have replaced the idea of citizens entitled to rights and services with 'paying clients' consuming commodities (Barchiesi, 2007; Khan, 2000). This semantic and discursive shift has been accompanied by the introduction of punitive measures such as cutting off water and electricity services when people fail to pay and forcing communities to install pre-paid water and electricity meters (Desai, 2002).

Furthermore, urban governance in Johannesburg has become increasingly business-oriented, with the formation of specialist agencies who work alongside local government to manage the business affairs, developmental needs and rejuvenation of the city. Organisations such as the Central Johannesburg Partnership (CJP), Johannesburg Property Company (JPC) and Johannesburg

Development Agency (JDA) are prominent in setting the direction and priorities of the city and serve to ensure cooperation between local government and the business sector (Lipietz, 2008). For instance, the JDA was established expressly for the purpose of encouraging partnerships between the City of Johannesburg and the private sector and encouraging businesses to invest in the city (Johannesburg Development Agency, 2010). It focuses on 'greenfield' projects designed to boost the image and attractiveness of the city (Viruly et al., 2010). This creates a situation in which regeneration projects prioritise capital and investors above the needs of ordinary citizens and low-income communities (Winkler, 2009).

Another clear, urban-level manifestation of the neoliberal, pro-business agenda has been the rise of City Improvement Districts (CIDs). CIDs are local variations of the Business Improvement Districts (BIDs) which have come to prominence in many Anglo-American cities. The adoption of this model in South Africa is symptomatic of the increasing authority of 'policy entrepreneurs,' who circulate globally and influence government agencies in policy formation and implementation (Peyroux, 2012, p. 178). CIDs are self-contained management districts which aim to improve predetermined geographic areas of the city by placing their day-to-day running and maintenance in the hands of independent, private boards. They are established when the majority of property owners and businesses within a circumscribed area agree to set up a non-profit management company. The company then charges every property owner in the defined area a levy, which is used to pay for top-up maintenance, cleaning and security services (Bénit-Gbaffou et al., 2012). Recent challenges to CID legislation in South Africa have made additional levies voluntary, but the model continues to be influential. CIDs have generally been successful urban management interventions and areas where they have been established have come to enjoy enhanced cleanliness and safety (Berg, 2004).

Significantly, however, CIDs are also increasingly criticised for the ways in which they fragment the urban landscape, introduce differential access to resources and services and pursue revanchist forms of urban management (Berg, 2010; Paasche et al., 2014; Peyroux, 2006). They also concentrate a great deal of decision-making and urban governance power in property owners' hands. Whilst the management boards established to run CIDs in theory consist of partnerships between residents, local government and property owners/the business sector, in practice, since voting rights are determined by the amount of property individuals own in the area and the amount contributed through levies, the more wealthy and powerful members come to have control (Bénit-Gbaffou, 2008). In Johannesburg, 'the City of Johannesburg openly acknowledges that it seldom participates in the central CID's management board' (Bénit-Gbaffou et al., 2008, p. 710), creating a situation in which the demands of the business sector are prioritised and they are given increased decision-making powers and control over space.

Developmental imperatives

The neoliberal policy and governance agendas have meant that, not only have the inequalities and injustices of the past endured, but new disparities and forms of deprivation have arisen in the post-apartheid period. Unemployment levels have increased dramatically as industry has stagnated and employment has become increasingly casualised. The white community continues to enjoy the largest share of wealth and income and also enjoys the highest living standards (Everatt, 2014). Although a small black elite and sizeable middle class have emerged, social mobility for some has been coupled with worsening economic prospects and quality of life for the majority of the country's citizens (Mushongera, 2017; Mushongera et al., 2017), leaving South Africa as one of the most unequal societies on the planet.

However, stress on the failings of the post-apartheid period can obscure many of the substantial gains and improvements which have been made since the end of apartheid. Particularly, it is worth highlighting how millions have been lifted out of poverty through government support and new access to employment and education opportunities (Seekings and Nattrass, 2005). The state has also suc-ceeded in constructing new housing for the poorest members of society, provid-ing homes for upwards of 11 million people (Tissington, 2009). Whilst pockets of deprivation and lack of necessities such as electricity, infrastructure and plumbing endure, these are the exception, rather than the norm. Furthermore, a substantial, progressive body of human rights has been codified in South African law, ensuring that all citizens, in theory at least, enjoy equal access to the law, education, basic needs such as housing, water and electricity, and are free to express their political opinions and sexual preferences. Although these steps have not alleviated the high levels of inequality, poverty and discrimination, they are positive developments in a country emerging from the shadows of racist author-itarian rule.

Constant depictions of post-apartheid policy and urban governance as 'neolib-eral' also fail to recognise the array of innovations and practices that are simultaneously driving urban change (Harrison et al., 2014; Miraftab, 2007; Mosselson, 2017). For example, Houghton (2013) shows how the imperatives of global competitiveness and economic growth are entangled with the pursuit of post-apartheid redress and economic redistribution in the city of Durban's urban development strategies and projects. Similarly, Sihlongonyane (2015) demonstrates how Johannesburg's city development and branding strategies, whilst focused on commercial growth, are also deeply invested in attempts to promote new articulations of cosmopolitan but distinctly African urban identities, showcase 'black excellence' and disprove prevalent associations of Africa with despair, disfunction and governance failures. Furthermore, whilst some of the urban renewal policies in Johannesburg are indeed instantiations of neoliberal ideologies and practices, it needs to be acknowledged that these are constantly played out whilst local government tries simultaneously to address legacies and daily realities of social exclusion and widespread poverty (Parnell and Robinson, 2006, 2012). Whilst neoliberalism and rule by market forces are prevalent tropes in the post-apartheid period, struggles for equality and liberation are seldom absent from South African discourse and daily life (van Holdt, 2012a).

Therefore, whilst government policies and programmes take on increasingly neoliberal forms, they retain commitments to redistributive and socially progressive goals. The South African state has also increased its spending on social welfare and support, and is today one of the largest middle-income welfare states in the world. Even with the increased neoliberal approach to governance and socio-economic life,

> the provision of services and infrastructure to meet the basic needs of historically disadvantaged populations is a widely accepted priority of the post-apartheid government. Their provision, with that of housing, is a major if not the major fiscal commitment to anti-poverty activity in Johannesburg.
>
> Beall et al. (2000, p. 114)

Reflecting these ambitions, the latest iteration of inner-city revitalisation strategies, the Inner City Transformation Roadmap, whilst maintaining the key pro-growth commitments spelled out in the Inner Charter, emphasises the need for local government to find ways to address 'the ongoing need for accommodation for very poor residents and newcomers to the city' and highlights poverty and housing shortages as 'the most critical issues in the inner city' (City of Johannesburg, 2013, p. 6).

The policies and political programmes adopted in the post-apartheid period therefore cannot be simply classified or made to fit into one overriding narrative. Attention needs to be paid to the ways in which alternative discourses, new imaginations and possibilities exist alongside powerful structures and inequalities, and how the city itself is a field through which these competitions are played out (Isin, 2008). The different trajectories and policy programmes which have shaped Johannesburg in the contemporary period serve as powerful reminders and evidence that the post-apartheid social order, particularly as it is expressed and lived through cities, remains uncertain, hybridised and in-process.

Awareness of this uncertainty and hybridity sits at the heart of my analysis of dynamics in the inner-city. My contention throughout is that diverse strands and pressures push and pull urban societies in different directions and that single-sided causal explanations and evaluations should be avoided in favour of more rounded, context-sensitive and nuanced critiques. In making my case, I draw on the notion of simultaneity, as articulated by Henri Lefebvre. Simultaneity, as a concept and heuristic device, is central to Lefebvre's reading of cities. He uses it to emphasise the vibrant, contested and always multiple and shifting nature of the urban and to underscore how cities are spaces of multiplicity and becoming (Schmid, 2008). At the same time, the urban is also engendered and dominated by capitalism and its mode of production (Prigge, 2008). Hence cities are simultaneously spaces of creativity, difference and emergence, whilst also being spaces of domination, oppression and the crystallisation of inequality (Lefebvre, 2003).

Cities as works in progress

Global trends?

The trends which define the bulk of post-apartheid urban scholarship are reflected in the debates which occupy the attention of many international urban scholars at present. Work on the restructuring of urban space under neoliberal governance and the ever-expanding scope of gentrification sit on one side of the spectrum. In these narratives, generalised urban processes are discernible, and bring diverse locations around the globe together in shared analytic and experiential fields (Brenner and Schmid, 2015; Brenner and Theodore, 2005; Peck, 2015). An exemplar of this view is Smith's claim that gentrification has been 'generalized' around the world and is a key component of a global strategy of neoliberal capitalism, which has led to a 'convergence between urban experiences in the larger cities of what used to be called First and Third Worlds' (Smith, 2002, p. 440). Whilst the language scholars use today has progressed and more nuanced approaches to studying gentrification and the ways it travels have appeared, efforts to tie diverse and disparate places and processes into over-arching, dominant narratives endure.

Examples of this include the narratives about 'planetary urbanisation' (Brenner and Schmid, 2015; Merrifield, 2013) and planetary or global gentrification(s) (Atkinson and Bridge, 2005; Lees et al., 2015). The former attempt to depict a world in which all spaces and locations are brought into the fold of capitalist urbanisation, and the urban itself becomes a universal category, with gentrification central to this process. The latter has emerged as an in-depth attempt to demonstrate how gentrification is both an endogenous and exogenous urban phenomenon, which has travelled across North–South, West–East divides, and also sprung up organically within different urban settings. It is, therefore, an attempt to exercise a comparative imagination and engage in learning that draws on different academic cultures and vocabularies, as well as practical case-studies and experiences (Lees, 2012). However, whilst laudable in its goals, the idea of planetary gentrification maintains at its core a sense that all urban change is predictable, fits under the same rubric, and can be criticised or dismissed with the same critical term. Thus, whilst the authors of *Planetary Gentrification* claim to be open to learning from and across different contexts and stretching conceptual imaginations (Lees et al., 2016), they also maintain a rigid definition and approach to urban redevelopment and revitalisation. For example, they utilise Clark's definition of gentrification which, as they point out, 'is not tied to the experience of a particular city and a particular time' (Lees et al., 2016, p. 12). It therefore falls into the trap of many Eurocentric, colonial thought-patterns in presuming to be thinking from 'nowhere,' or claiming temporally bounded, place-specific experiences as universal. The authors make use of his definition which declares:

> Gentrification is a process involving a change in the population of land-users such that the new users are of a higher socio-economic status than the

previous users, together with an associated change in the built environment through a reinvestment in fixed capital... It does not matter where, it does not matter when. Any process of change fitting this description is, to my understanding, gentrification.

(Clarke, 2005, p. 258, cited in Lees et al., 2016, p. 12)

I do not wish to take issue with the idea of gentrification, or the understanding that it both travels and has emerged in numerous places around the globe, with deleterious effects on local populations. What I take issue with is the presumed *inevitability* of the process, and shared conclusions which authors, operating with a singular definition, reach.

My sympathies are more closely aligned with the schools of thought which are broadly recognised as comparative and postcolonial. Scholars working within these traditions attempt to draw attention away from implied linear or universalising narratives about urban society, and rather focus on the emergence, inventiveness, uncertainty and diversity which characterise different societies, particularly those in postcolonial contexts. Myers (2011) identifies two broad categories in urban scholarship on Africa: one he terms the 'materialist school,' which is concerned predominantly with neoliberal governance across African cities, and the second he describes as broadly post-structuralist. The latter is less concerned with broad-scale political and macro-economic trends, and focuses more on seeing African cities 'as works in progress' (Simone, 2004b, p. 1) and examining the practices through which urban life is made and sustained in precarious, uncertain environments. Myers holds up the work of AbdouMaliq Simone (for example see Simone, 2001, 2004a, 2008, 2010) and de Boeck and Plissart (see Plissart and de Boeck, 2006) as exemplars of this approach. Drawing on his own research and that of others, he embraces the improvisational slant of the poststructuralists, but uses it to focus on the 'thickness and messiness' (Myers, 2011, p. 125) of practical attempts to govern and deliver services in contemporary urban African settings. Through this body of work, we have come to recognise that rather than a homogenous story of deprivation, inequality and suffering induced by neoliberal policies, African cities are sites where radically experimental and inventive politics and practices unfold (Pieterse, 2011). In these analytic approaches and the case studies which they draw on, the means for enacting urban processes are uncertain, and people have to rely on a range of materials, resources, networks, myths and belief systems in order to secure and reproduce urban life.[7] Importantly, if the acts of governing are themselves uncertain and improvised, then it follows that the outcomes are too.

Following those who focus on the situated, in-process actions which make postcolonial cities, I attempt to develop a theoretical perspective which is attuned to the pragmatic, reflexive and evolving practices behind urban regeneration activities in inner-city Johannesburg. Rather than seeing the process and those at the forefront of it as operating with fixed, pre-determined agendas and logics, which give rise to predictable outcomes, I prefer to expand on the uncertain and intertwined relationship between practices and places, and to foreground the ways in which practices, even on the parts of powerful actors such as property developers and private security personnel, adapt to suit the contexts in which they unfold. On the one hand the adaptations which I describe in later chapters serve to make commercial enterprises and security practices more agile and better suited to taming inner-city environments, but they also come to reflect the logics and socio-spatial realities, including poverty, racial transition and informality, which characterise the inner-city. As such, they signify the ways in which practices and outlooks adapt and change in the face of complex, multi-faceted and diverse urban contexts.

Urban regeneration as a spatial praxis

I turn to the work of Lefebvre and Pierre Bourdieu to develop my perspective. Although there aren't many examples where their work has been combined or engaged with simultaneously, Lefebvre and Bourdieu have several common themes running through their work (although there are also several significant differences). At its heart, Bourdieu's social theory has a concern with social order and class hierarchies and the ways they are reproduced in society (van Holdt, 2013). Similarly, Lefebvre is concerned with how cities and urban spaces are central to the production and reproduction of the capitalist spatial and social order (Merrifield, 2006). Both are intrigued by and help us understand how people live under systems of domination and come to reproduce these, whilst simultaneously retaining the capacities for individual action and expression. They also demonstrate how, whilst people do remain creative, agentful and able to shape the worlds around them, their actions also serve to reproduce social order and systems of stratification. Returning to the particularities of the South African context, their work can help us understand the tensions between enduring patterns of racism, inequality, constricting neoliberal impulses and the optimism and progressive politics of the post-apartheid period.

Bourdieu's social theory regards domination and social stratification as reinforced and reproduced through social activities and institutions. His analytic account focuses on people's decisions and how all of these not only express the effects of social domination, but also reproduce the hierarchies and forms of stratification which capitalist society rests on (Bourdieu, 1984, 1990, 2005a). His is thus a theory which sees domination played out in individual's ordinary lives and experiences, and which therefore centres on agency and activity. It is a multi-faceted approach to questions of domination and the relationship between social structure and agency, and helps arrive at a nuanced understanding of how social orders are constituted and reproduced (Kelly and Lusis, 2006). Lefebvre too focuses attention on daily life and lived reality. He treats it as a site of both domination and rebellion. Daily life in urban spaces is produced and overdetermined by the capitalist mode of production, and this system moulds people's identities, social relationships, forms of habitation and daily routines (Kipfer, 2008; Lefebvre, 1991; Prigge, 2008; Stanek, 2011). At the same time, however, Lefebvre calls attention to the fact that spaces are constantly negotiated and contested (Goonewardena, 2008). As is clear from the changes which have occurred in central Johannesburg, the ways in which people inhabit spaces and come to make lives and dwellings for themselves within them shape these spaces and inscribe them with meanings which can challenge or refute the dominant narratives and schemas through which they have been produced. Both Bourdieu and Lefebvre's theoretical frameworks therefore draw attention to the ways in which daily life is complex and open to a variety of readings and processes.

An approach which focuses on the relationship between daily life, reproduction and social order is suitable for exploring urban regeneration, as this too is a form of creating and maintaining social order, or at least attempting to do so. Urban regeneration is about the transformation of cityscapes, not only their built form, but also in terms of how they are experienced, who gets to use and inhabit them, and the cultural meanings they are inscribed with (Bridge, 2001; Ley, 2003; Vicario and Martinez Monje, 2005; Zukin, 1998). In Johannesburg, inner-city regeneration is frequently a process which seeks to bring order into decayed and chaotic spaces, to uplift areas which have been crippled by capital flight, to attract investment and to cultivate particular styles of living and ways of being urban (Gaule, 2005; Murray, 2008). It is thus deeply invested in efforts to produce social order and to reproduce it through the spaces which are created and the ways they are used. However, as earlier discussion in this chapter pointed out, it is uncertain what this social order represents at present; it has features of conflict and exclusion, as seen in the spread of CIDs and a regeneration agenda which centres on attracting private investment, but it also incorporates transformative and democratic features and possibilities. Paying close attention to the ways in which the social order is emerging and how it is being reproduced, as well as how it is contested and subverted, allows us to, ultimately, analyse what it signifies – an exclusionary, constraining order, a progressive, transformative one, or a hybrid combination of both.

In urban geography, Bourdieu's work has most frequently been associated with demand-side, culture-based explanations of gentrification (see Butler and Robson, 2003; Ley, 2003; Savage, 2011). More recently, some scholars have begun drawing on his ideas in exploring the spatial dimensions of social distinction, social reproduction and domination. Rérat and Lees (2011) have linked spatial capital to locational advantage, and use the term to show how new-build gentrification in European border areas enables the gentrifying class to claim mobility and lifestyle advantages which allow them to reproduce the work and commuting practices which secure their social reproduction, and how the beneficiaries of investments in the built environment and new housing developments are most frequently those already in privileged economic and social positions. They therefore draw attention to the role access to urban space plays in reproducing economic and social hierarchies. These insights are key to understanding inequality in South Africa. Apartheid spatial patterns endure at present, with the majority of poor urban residents living in townships and informal settlements located on the peripheries of large cities. Their access to employment opportunities, resources and amenities is restricted, and they are faced with long, expensive commutes to reach places of work and education (Todes, 2014; Mokonyama and Mubiwa, 2014). Consequently, urban regeneration which allows people to access housing in central areas is vital to meeting some of the developmental challenges of the contemporary period and can provide lower-income households with improved forms of spatial and economic capital. However, the tensions of a market-driven urban agenda conflict with these imperatives, and make providing housing which can cater to poor members of society difficult, demonstrating the conflicting agendas which urban regeneration is driven by.

Taking the notion of domination and reproduction further, Marom (2014) has applied Bourdieu's understanding of distinction to processes of urban development, and uses the notion of 'spatial distinction' to highlight how processes of spatial segregation are, in part, driven by the differential forms of classification and prestige which circulate in society in various periods of time, and which actively contribute to the ways in which cities are built, territories are labelled and populations are divided within them. Working in the context of Israeli colonialism, he develops insights about the relationships between race, segregation, planning and (under)development which are equally relevant for South Africa, where associations between race, place and discrimination are rife. For instance, financial institutions' red-lining of inner-city Johannesburg is based more on racist perceptions than economic reality (Huchzermeyer and Haferburg, 2017; Murray, 2008), and demonstrates clearly how cultural preferences and hierarchies of perception are inscribed into the built environment, and come to reproduce patterns of discrimination and deprivation. Centner (2008) follows a similar approach to Marom in highlighting the intertwining between social distinction and the production of space, and uses the term 'spatial capital' to explicitly focus on how dominant groups, in his case dot-com entrepreneurs, are able to, literally, *take and make place*, as their consumption habits, dominant economic position and resulting social prestige come to shape physical locations and, in so doing, displace other social groups and delegitimise their claims to space. Struggles over urban space in South Africa, particularly in Cape Town (see Didier et al., 2012; Teppo and Millstein, 2015) and Johannesburg's pockets of gentrification (see Nevin, 2014; Walsh, 2013), certainly speak to these processes too, and demonstrate how efforts to cater to economically and culturally dominant groups' consumption preferences remake physical spaces, leaving others feeling out of place, or even physically displaced (Wilhelm-Solomon, 2016).

These perspectives are all vital and are discussed in more detail later in the book. What I will highlight, and which is arguably absent from the existing body of work, is a sense of the *dynamism* of space and spatial practice. By drawing on Lefebvre's understanding of space as productive, and Bourdieu's understanding of habitus as socially learned, embodied and reproduced through daily life, I aim to highlight how space is not just acted on and produced by dominant actors, but that being able to acquire spatial capital and, in so doing, produce space, means acquiring a habitus which reflects and responds to the contingencies and multiple, dynamic realities of lived space. I therefore use spatial capital, following Centner (2008), as the ability to take and make place, but also as the ability to successfully navigate, inhabit and engage with space. This ability is therefore something which is subsumed and enacted through one's habitus – i.e. the set of socially and spatially learnt forms of apprehension, dispositions and propensities to act (Bourdieu, 2005b). Through this lens, I hope to shed new light on thinking about the potential outcomes of urban revitalisation processes, as well as some of the people behind them. I also hope to offer up a novel approach to thinking through space and social action, and their mutually reinforcing relationship.

I do not present a programmatic engagement with or reading of Bourdieu and Lefebvre's theories. Nor is this work a homage to them, both of whom had next to nothing to say about contexts such as the one I am working in in their own universalising theories about the world. In particular, recognition of the role racism plays in fostering forms of domination and urban inequality is absent from their work. In South Africa, race has to remain at the forefront of how we think about patterns and processes of inequality, discrimination, deprivation, as well as assertions of identities, ideals of democracy and aspirations. My work, then, is intended as a creative engagement with theory and an attempt to ground and adapt theories originating elsewhere, so that they may be used to help make sense of a new locality and subject area. In doing so, I hope to engage in the type of agile and flexible theorisation which Robinson (2016) calls for, and also provide a reminder that Southern contexts can be sites through which Eurocentric ideas can be rethought, questioned and altered, and from which new theories can be generated (Comaroff and Comaroff, 2012; Roy, 2009; van Holdt, 2012b).

Structure of the book

The proceeding chapters draw on the debates, conceptual frameworks and epistemological approaches outlined above, and present a detailed, multi-faceted study of the regeneration process underway. The account which follows engages with both the simultaneity and diverse imperatives shaping the way regeneration is taking place, as well as the interplay between spatial dynamics, legacies and contemporary realities of racial discrimination and disadvantage, agency and everyday life. The aim of the book is to demonstrate the ways in which the regeneration process is reproducing policies and urban governance practices which follow neoliberal logic, but to also underscore the uncertainty, openness and positive outcomes which have also been engendered. My overall goal is to open our readings of urban processes and emphasise the multiplicity of agendas, ideologies and practices which go into shaping urban interventions, and which exist simultaneously in a variety of settings. The book therefore stands as a provocation which urges scholars concerned with other projects or locations to recognise competing logics and vernacular specificity, but without abandoning comparative ambitions and practices either.

The chapters which follow draw on insights and information gathered over a lengthy period of fieldwork in Johannesburg. Research entailed qualitative interviews with a variety of actors involved in inner-city regeneration and housing provision, including housing providers, property developers, private security personnel, building managers,[8] government officials and civil society groups. In addition, interviews were conducted with 57 tenants spread across seven renovated buildings (two social housing and five for-profit). Due to the demographics of the tenant populations in these buildings, all participants were black (the majority being Black African) and the majority were South African citizens. However, several foreign nationals from countries including Zimbabwe, Nigeria and the Democratic Republic of Congo were also included in the sample. Interviews were supplemented by ethnographic observation, including attending community events, planning meetings for these events, security shifts and street patrols with the Hillbrow Community Policing Forum.

Of course, it is not easy to gain access and insights into people's experiences, particularly when the researcher is socially distant from those whom s/he is studying. Whilst fully aware that it is not possible to simply enter and understand other people's worlds, I tried to familiarise myself as much as possible with the inner-city environment, and spent six months living in an apartment in the area, volunteered at various community events and spent time socialising in a local park in order to gain more acquaintances in and direct experiences of the neighbourhoods I was studying. These methods are in some ways superficial and cannot overcome the barriers which social biographies inscribe, but did assist in shaping my insights into the area and the different issues and dynamics which characterise it. Because I approached the area and the people who are living in it from a distance, I had to remain careful to not impose a preconceived set of perceptions onto them, and to do my utmost to build a narrative inductively out of the descriptions, details and experiences which people shared with me. These experiences are multi-faceted and defy explication into one account. Hence this book revolves around contradictory details and assembles them together into a picture which tries to do justice to and accurately reflect the complexity and ambiguity inherent in people's everyday lives and the practices they engage in.

Map 1.2 depicts the inner-city areas and different residential buildings in which research was conducted, and Table 1.1 specifies the types of housing each building provides.

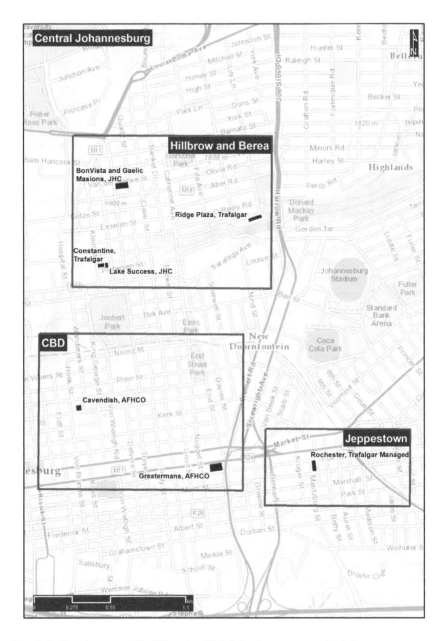

Map 1.2 Overview map of buildings in which interviews were conducted.

Table 1.1 Buildings in which interviews were conducted.

Building Name	Housing Type	Number of Tenants Interviewed	Location
Cavendish Court	For-profit	5	CBD
Constantine	For-profit	4	Hillbrow
Gaelic Mansions	Social	18	Hillbrow
Greatermans	For-profit	6	CBD
Lake Success	Social	9	Hillbrow
Ridge Plaza	For-profit	6	Berea
Rochester	For-profit	9	Jeppestown

Chapter 2, 'An overburdened process: the competing agendas, imperatives and outcomes of inner-city regeneration,' is situated within debates between neoliberal urban governance and developmentally oriented strategies. It engages with the various currents and imperatives which define governance and policy formation in the post-apartheid period, and examines the effects these have on the way urban regeneration has been formulated and financed. Drawing on interviews with several representatives of institutions providing finance for housing development in the inner-city, as well as local government representatives and civil society organisations active in the area, it demonstrates that urban regeneration is attempting to fulfil multiple, competing agendas. On the one hand, the process is firmly embedded within a market-based paradigm, with profiteering and augmenting the value of inner-city property at the heart of financing and housing provision strategies. Yet, there are also discernible developmental goals underlying the practices of finance agencies and policy-makers. Whilst laudable, the latter agenda is not easily reconciled with the market-based approach, leaving regeneration as an ambiguous process defined by contradictions and competing imperatives and outcomes. The chapter therefore shows that governance and policy-formulation is not an 'either-or' situation, where an approach or practice is either neoliberal or developmental, but that there is a simultaneous entanglement of both (broadly defined) agendas.

The following chapter, 'The contradictory praxis of regeneration,' demonstrates the effects that the policy and financing frameworks have on the practices of housing providers and property developers. It thus moves the discussion from the policy arena to focus on the practical activities which go into urban regeneration. Drawing on Bourdieu's understanding of social action taking place in fields, it demonstrates how urban regeneration is a distinct field, which inculcates and valorises specific forms of capital, dispositions and practices. The chapter presents a sociology of housing developers, a powerful but generally understudied set of actors, and demonstrates that the habitus and forms of distinction which underpin their worldviews and actions reflect the contradictions and competing agendas outlined in the previous chapter. The group of developers in my study are predominantly white, and therefore part of a society which has generally been hostile to urban change in South Africa (Ballard, 2010; Popke and Ballard, 2004). However, the policy framework which underpins the regeneration process, the general political atmosphere in the country and the social-spatial realities of the inner-city inculcate more developmentally oriented and socially sensitive dispositions and practices amongst this group. They therefore occupy dualistic positions, valorising social commitment and developmental practices, whilst simultaneously enacting and reinforcing neoliberal values and the logics of the market. In this chapter I expand on theorisations of 'spatial capital' and 'spatial habitus,' and demonstrate how housing providers' practices are reflections and outcomes of the particular space they are located in. It will become clear that whilst they are dominant actors, they do not only *make* place, but that the place itself shapes and influences their actions and determines the types of practices they can engage in. Thus, attention is turned to the ways in which urban regeneration, rather than being a settled, inevitable process, is, in fact, a lived praxis requiring adaptability and capacities to adjust dispositions and practices to suit the specific environments in which it takes place. Although reflexive and attuned to the sensitivities of the inner-city, this group's dominance also raises important questions about power and patterns of ownership, and shows how white people continue to be disproportionally economically, socially and culturally dominant in South Africa. This becomes particularly clear when the perspectives of some black developers are voiced in the latter stages of the chapter, as well as in later chapters when tenants' perspectives and experiences are contrasted with housing developers' powers to make place and exercise spatial capital.

Following the exploration of urban regeneration as a lived, practical activity, the next chapter, 'Urban management and security: private policing, atmospheres of control and everyday practices,' focuses on the management and policing practices which have accompanied processes of urban upgrading in the inner-city neighbourhoods Hillbrow and Berea. Drawing on ethnographic research, participant observation and interviews with a range of housing management and private security personnel, the chapter discusses the everyday activities which bring order to Hillbrow, but which also target certain populations that are regarded as threatening or dangerous. Whereas most urban scholarship in South Africa has focused on spectacular modes of policing and displays of force (Cooper-Knock, 2016), particularly police-led crackdowns on informal trading and foreign migrants (for example Clarno and Murray, 2013; McMichael, 2015), I argue that these are only episodic interventions, and that urban security is produced through more mundane, routine, but equally (or perhaps even more) powerful forms of policing. Continuing to emphasise the hybridity and contradictory agendas which shape regeneration efforts in the inner-city, the chapter also demonstrates that the regime of order and security which has been created in Hillbrow attempts to be inclusive and accommodate some of the practices and populations that have been targeted elsewhere (particularly informal traders), but is also exclusionary and revanchist when it comes to other groups who are present in the area (notably homeless young men and residents of derelict buildings).

The penultimate set of chapters looks at the regeneration process from the perspectives of people who have been able to access social or affordable housing. Firstly, 'Ambiguous experiences of regeneration: spatial capital, agency and living in-between' details the ways in which residents have benefited from centrally located housing, and also highlights some of the new forms of associational life, friendships and forms of belonging and attachment which are arising in the inner-city. The chapter therefore underscores the transformational potential and outcomes of regeneration and housing provision for lower-income groups. Again, though, experiences and outcomes can never be generalised or reduced to singular accounts. The chapter therefore also documents some of the difficulties which residents encounter, and shows how they live in the inner-city not out of choice but because of the limited alternative affordable options available to them. Tenants' accounts show how they have to contend with a range of unpleasant, trying circumstances in the inner-city, and generally aspire to live elsewhere. However, the fragmented, unequal geography of the city means that there are few other options which they regard as suitable or can afford. They therefore make difficult compromises and live in in-between states, hoping to live elsewhere but recognising that it is unlikely they will be able to do so. They therefore adjust their dispositions, resigning themselves to getting by in far-from-ideal circumstances and concentrating on making lives in the inner-city. In these ways, they demonstrate how their positions in the city's social hierarchy are absorbed into their habitus and become expressions of their identities and experiences of post-apartheid urban citizenship. The chapter therefore shows that, whilst housing developers utilise their spatial capital to remake the inner-city, lower-income residents are placed in positions where they accept their situations and dismiss their abilities to change urban landscapes. It therefore becomes clear how experiences in space both reflect and reproduce social hierarchies and patterns of domination and inequality.

Next, 'The space that regeneration makes: regulation, security and everyday life,' turns attention to experiences inside residential buildings themselves. It shows how the regime of property-based development and the commercial imperatives underpinning the regeneration process create strict atmospheres and forms of surveillance and regulation inside residential buildings. Although giving tenants peace of mind, these systems ensure that housing companies are in dominant positions and tenants cannot question or challenge living arrangements. They also frequently infringe on rights and protections granted to tenants, and show how commercial imperatives are placed ahead of citizens' rights and forms of communal life. Again, the inner-city's social hierarchy is reinforced, and tenants are placed in subordinate positions to developers and property owners. At the same time, however, some experiences of communal solidarity and mutual care and concern are recounted. These instances demonstrate how tenants make lives for themselves and others in the inner-city, and, in doing so, create new meanings for and experiences of the post-apartheid city. The chapter therefore draws attention to the everyday as a site of agency, where forms of appropriation, solidarity and the politics of change play out.

Lastly, 'Towards a vernacular theorisation of urban change' sums up the key arguments and insights from the previous chapters and takes the theoretical perspectives and learnings forward. It articulates some of the key recommendations or findings which my research has yielded, and also makes broader theoretical points about understanding processes of urban change and moving between specific cases and more generalisable, comparative trends.

Notes

1 Racial terms are used frequently in this book as they reflect both a historical legacy and pervasive reality of racial classification in South Africa. Whilst racial categorisations are social constructions and have generally been used in harmful, problematic ways, they continue to structure many people's lives, interactions and identities in South Africa, and have ongoing salience and social relevance. Following census definitions, 'Black African' refers to indigenous African people, 'coloured' refers to people of mixed race, and 'black' to all groupings other than whites.

2 A cornerstone of colonial and apartheid ideologies was the idea that white people were the heirs to European civilisation, and their mission was to bring religion, modernity and progress to Africa. 'European' urban spaces were contrasted with the rural homelands which black people were exiled to. Whilst modern Dutch and Roman law were used to govern South Africa's cities and white spaces, the black population in homelands was subject to rule by a distorted form of traditional law (Mamdani, 1996). Hence, cities were entwined with and the spatial manifestations of the 'bifurcated' system which colonial and apartheid rule was based on. 'European' was also an everyday signifier and marker of difference, and was used as a synonym for 'white.' For instance, apartheid segregation was enforced by designating certain public spaces as solely for European's use.

3 www.news24.com/SouthAfrica/News/live-wits-fees-protest-turning-violent-20161010.

4 Influx control laws were introduced in several stages during the colonial and apartheid eras and were used to regulate the presence of black people in areas designated as 'white' or 'European.' Under these laws, black people were forced to carry passes which identified them and their employment status and determined whether they had been granted permission to reside (always temporarily) in white urban areas. The Group Areas Act was first introduced in 1950 and it formed the basis of the country's segregated urban landscape. It divided the country's cities into different residential areas which were determined by race and forced people to reside in these areas. Under this law there were several neighbourhood demolitions and forced removals, and combined with influx control and Pass Laws, urban spaces were made inaccessible and hostile to black people (Bonner, 1995; Morris, 1999a; van Onselen, 2001).

5 In contrast, the Reconstruction and Development Programme (RDP), the first economic policy of the post-apartheid era, emphasised direct government intervention in the economy and the expansion of public participation and welfare services as the key features through which socio-economic inequalities and growth could be addressed.

6 A large body of scholarship has critiqued the largely deleterious effects these governance interventions have had: for example Hart (2002) and Naidoo (2007) draw attention to neoliberal governing mechanisms, such as pre-paid water and electricity meters, at community level; examples of how restructuring and the increased flexibility of employment have made work in the retail and manufacturing industries more precarious and less financially rewarding are found in the work of several sociologists (see Bezuidenhout et al., 2007; Kenny, 2004; Kenny and Webster, 1998; Mosoetsa, 2011 amongst others); and the overall trends of increasing inequality, unemployment and social tensions are captured by political economists and geographers such as Habib (2013), Murray (2011) and Bond (2000).

7 Alongside the 'poststructuralist' urban studies school, the growing body of ethno-graphic work on state practices in Africa and other postcolonial settings provides rich, important detail about the ways in which officials go about their everyday tasks, and have to cope with opaque systems, resource shortages and transient, indecipherable populations when doing so (see Bierschenk and De Sardan, 2014; Hoag, 2010; van Holdt, 2010). This work, alongside research focusing on policy-makers and urban planners (for example Lipietz, 2004; Robinson, 2015), also shows the invaluable contribution which state actors can make in delivering services, providing basic necessities and accomplishing meaningful change, as well as some of the obstacles which they face in attempting to do so.

8 Building managers, also referred to by some companies as housing supervisors, are employees of housing companies who live inside residential buildings and are responsible for day-to-day management and maintenance, security arrangements and relationships with tenants. In looking after basic maintenance, they play similar roles to property caretakers, but they also fulfil more professional, administrative tasks such as handling lease agreements with tenants, ensuring rent is collected and disciplining tenants if needs be.

References

Atkinson, R., Bridge, G. (Eds.), 2005. *Gentrification in A Global Context: The New Urban Colonialism*. Routledge. London.

Back, L., 2015. Why everyday life matters: class, community and making life liveable. *Sociology*. 49, 820–836.

Ballard, R., 2010. 'Slaughter in the suburbs': livestock slaughter and race in post-apartheid cities. *Ethnic and Racial Studies*. 33, 1069–1087.

Barchiesi, F., 2007. Privatization and the historical trajectory of 'social movement unionism': a case study of municipal workers in Johannesburg, South Africa. *International Labor and Working-Class History*. 71, 50–69.

Beall, J., Crankshaw, O., Parnell, S., 2000. Local government, poverty reduction and inequality in Johannesburg. *Environment and Urbanization*. 12, 107–122.

Beavon, K., 2004. *Johannesburg: The Making and Shaping of the City*. University of South Africa Press. Pretoria.

Bénit-Gbaffou, C., 2008. Unbundled security services and urban fragmentation in post-apartheid Johannesburg. *Geoforum*. 39, 1933–1950.

Bénit-Gbaffou, C., Didier, S., Morange, M., 2008. Communities, the private sector, and the state: contested forms of security governance in Cape Town and Johannesburg. *Urban Affairs Review*. 43, 691–717.

Bénit-Gbaffou, C., Didier, S., Peyroux, E., 2012. Circulation of security models in southern African cities: between neoliberal encroachment and local power dynamics. *International Journal of Urban and Regional Research*. 36, 877–889.

Berg, J., 2004. Private policing in South Africa: the Cape Town City Improvement District – pluralisation in practice. *Society in Transition*. 35, 224–250.

Berg, J., 2010. Seeing like private security: evolving mentalities of public space protection in South Africa. *Criminology and Criminal Justice*. 10, 287–301.

Bezuidenhout, A., Khunou, G., Mosoetsa, S., Sutherland, K., Thoburn, J., 2007. Globalisation and poverty: impacts on households of employment and restructuring in the textiles industry of South Africa. *Journal of International Development*. 19, 545–565.

Bierschenk, T., De Sardan, J.P.O. (Eds.), 2014. *States at Work: Dynamics of African Bureaucracies*. Brill. Leiden.

Bond, P., 2000. *Elite Transition: From Apartheid to Neoliberalism in South Africa*. Pluto Press. Cape Town.

Bonner, P., 1995. African urbanisation on the rand between the 1930s and 1960s: its social character and political consequences. *Journal of Southern African Studies*. 21, 115–129.

Bourdieu, P., 1984. *Distinction: A Social Critique of the Judgement of Taste.* Harvard University Press. Cambridge, Massachusetts.

Bourdieu, P., 1990. *The Logic of Practice.* Stanford University Press. Stanford, California.

Bourdieu, P., 2005a. *The Social Structures of the Economy.* Polity Press. Cambridge.

Bourdieu, P., 2005b. Habitus, in: Hillier, J., Rooksby, E. (Eds.), *Habitus: A Sense of Place.* Ashgate. Aldershot, pp. 43–49.

Bremner, L., 2000. Reinventing the Johannesburg inner city. *Cities.* 17, 185–193.

Brenner, N., Schmid, C., 2015. Towards a new epistemology of the urban? *City.* 19, 151–182.

Brenner, N., Theodore, N., 2005. Neoliberalism and the urban condition. *City.* 9, 101–107.

Bridge, G., 2001. Bourdieu, rational action and the time–space strategy of gentrification. *Transactions of the Institute of British Geographers.* 26, 205–216.

Butler, T., Robson, G., 2003. Negotiating their way in: the middle classes, gentrification and the deployment of capital in a globalising metropolis. *Urban Studies.* 40, 1791–1809.

Centner, R., 2008. Places of privileged consumption practices: spatial capital, the dot-com habitus, and San Francisco's internet boom. *City and Community.* 7, 193–223.

Chari, S., 2017. The Blues and the Damned: (Black) life-that-survives capital and biopolitics. *Critical African Studies.* 9, 152–173.

Charlton, E., 2017. Melancholy mapping: a 'dispatcher's eye' and the locations of loss in Johannesburg. *Thesis Eleven.* 141, 14–30.

Charlton, S., 2009. Housing for the nation, the city and the household: competing rationalities as a constraint to reform? *Development Southern Africa.* 26, 301–315.

Charlton, S., Kihato, C., 2006. Reaching the poor? An analysis of the influences on the evolution of South Africa's housing programme, in: Pillay, U., Tomlinson, R., Du Toit, J. (Eds.), *Democracy and Delivery: Urban Policy in South Africa.* HSRC Press. Cape Town, pp. 252–282.

Chipkin, C.M., 1993. *Johannesburg Style: Architecture & Society, 1880s–1960s.* David Philip Publishers. Cape Town.

City of Johannesburg, 2007. *Inner City Regeneration Charter.* City Council. Johannesburg.

City of Johannesburg, 2013. *Inner City Transformation Road Map.* City Council. Johannesburg.

Clarno, A., 2013. Rescaling white space in post-apartheid Johannesburg. *Antipode.* 45, 1190–1212.

Clarno, A., Murray, M.J., 2013. Policing in Johannesburg after apartheid. *Social Dynamics.* 39, 210–227.

Comaroff, J., Comaroff, J.L., 2012. *Theory from the South: Or, How Euro-America Is Evolving toward Africa.* Paradigm Publishers. Boulder, Colorado.

Cooper-Knock, S.J., 2016. Behind closed gates: everyday policing in Durban, South Africa. *Africa.* 86, 98–121.

Crankshaw, O., White, C., 1995. Racial desegregation and inner city decay in Johannesburg. *International Journal of Urban and Regional Research.* 19, 622–638.

de Vries, F., 2008. Megamalls, generic city, in: Nuttall, S., Mbembe, A. (Eds.), *Johannesburg: The Elusive Metropolis.* Duke University Press. Durham, North Carolina, pp. 297–306.

Desai, A., 2002. *We Are the Poors: Community Struggles in Post-Apartheid South Africa.* Monthly Review Press. New York.

Didier, S., Peyroux, E., Morange, M., 2012. The spreading of the city improvement district model in Johannesburg and Cape Town: urban regeneration and the neoliberal agenda in South Africa. *International Journal of Urban and Regional Research*. 36, 915–935.

Dinath, Y., Patel, Y., Seedat, R., 2014. Footprints of Islam in Johannesburg, in: Harrison, P., Gotz, G., Todes, A., Wray, C. (Eds.), *Changing Space, Changing City: Johannesburg After Apartheid*. Wits University Press. Johannesburg, pp. 456–480.

Dirsuweit, T., Wafer, A., 2006. Scale, governance and the maintenance of privileged control: the case of road closures in Johannesburg's Northern Suburbs. *Urban Forum*. 17, 327–352.

Everatt, D., 2014. Poverty and inequality in the Gauteng city-region, in: Harrison, P., Gotz, G., Todes, A., Wray, C. (Eds.), *Changing Space, Changing City: Johannesburg After Apartheid*. Wits University Press. Johannesburg, pp. 63–82.

Gaule, S., 2005. Alternating currents of power: from colonial to post-apartheid spatial patterns in Newtown, Johannesburg. *Urban Studies*. 42, 2335–2361.

Goga, S., 2003. Property investors and decentralization: a case of false competition? in: Tomlinson, R., Beauregard, R.A., Bremner, L., Mangcu, X. (Eds.), *Emerging Johannesburg: Perspectives on the Postapartheid City*. Routledge. New York, pp. 71–84.

Goonewardena, K., 2008. Marxism and everyday life: on Henri Lefevbre, Guy Debord, and some others, in: Goonewardena, K., Kipfer, S., Milgrom, R., Schmid, C. (Eds.), *Space, Difference, Everyday Life: Reading Henri Lefebvre*. Routledge. New York, pp. 117–133.

Gotz, G., Simone, A., 2003. On belonging and becoming in African cities, in: Tomlinson, R., Beauregard, R.A., Bremner, L., Mangcu, X. (Eds.), *Emerging Johannesburg: Perspectives on the Postapartheid City*. Routledge. New York, pp. 123–147.

Gotz, G., Todes, A., 2014. Johannesburg's urban space economy, in: Harrison, P., Gotz, G., Todes, A., Wray, C. (Eds.), *Changing Space, Changing City: Johannesburg After Apartheid*. Wits University Press. Johannesburg, pp. 117–136.

Gumede, W.M., 2007. *Thabo Mbeki and the Battle for the Soul of the ANC*. Zebra Press. Cape Town.

Habib, A., 2013. *South Africa's Suspended Revolution: Hopes and Prospects*. Ohio University Press. Athens, Ohio.

Harrison, P., Gotz, G., Todes, A., Wray, C., 2014. Materialities, subjectivities and spatial transformation in Johannesburg, in: Harrison, P., Gotz, G., Todes, A., Wray, C. (Eds.), *Changing Space, Changing City: Johannesburg After Apartheid*. Wits University Press. Johannesburg, pp. 2–39.

Harrison, P., Harrison, K., 2014. Soweto: a study in socio-spatial differentiation, in: Harrison, P., Gotz, G., Todes, A., Wray, C. (Eds.), *Changing Space, Changing City: Johannesburg After Apartheid*. Wits University Press. Johannesburg, pp. 293–318.

Harrison, P., Huchzermeyer, M., Mayekiso, M. (Eds.), 2003. *Confronting Fragmentation: Housing and Urban Development in a Democratising Society*. University of Cape Town Press. Cape Town.

Hart, G.P., 2002. *Disabling Globalization: Places of Power in Post-Apartheid South Africa*. University of California Press. California.

Hlongwane, G., 2006. 'Reader, be assured this narrative is no fiction': the city and its discontents in Phaswane Mpe's *Welcome to Our Hillbrow*. *ARIEL*. 37, 69.

Hoag, C., 2010. The magic of the populace: an ethnography of illegibility in the South African immigration bureaucracy. *PoLAR: Political and Legal Anthropology Review*. 33, 6–25.

Houghton, J., 2013. Entanglement: the negotiation of urban development imperatives in Durban's public–private partnerships. *Urban Studies*. 50, 2791–2808.

HSRC, 2014. *The City of Johannesburg (COJ) Economic Overview: 2013. A Review of the State of the Economy and Other Key Indicators, Economic Performance and Development Programme (EPD)*. Human Sciences Research Council. South Africa.

Huchzermeyer, M., 2001. Housing for the poor? Negotiated housing policy in South Africa. *Habitat International*. 25, 303–331.

Huchzermeyer, M., 2014. Changing housing policy in South Africa, in: Bredenoord, J., Van Lindert, P., Smets, P. (Eds.), *Affordable Housing in the Urban Global South: Seeking Sustainable Solutions*. Routledge. London, pp. 336–348.

Huchzermeyer, M., Haferburg, C., 2017. Redlining or renewal? The space-based construction of decay and its contestation through local agency in Brixton, Johannesburg, in: Kirkness, P., Tije-Dra, A. (Eds.), *Negative Neighbourhood Reputation and Place Attachment: The Production and Contestation of Territorial Stigma*. Routledge. London, pp. 60–80.

Isin, E.F., 2008. The city as the site of the social, in: Isin, E.F. (Ed.), *Recasting the Social in Citizenship*. University of Toronto Press. Toronto, pp. 261–280.

Jinnah, Z., Rugunanan, P., 2016. Remaking religion, rethinking space: how South Asian and Somali migrants are transforming ethnically bound notions of Hinduism and Islam in Mayfair and Fordsburg, in: Wilhelm-Solomon, M., Nunez, L., Kankonde Bukasa, P., Malcomess, B. (Eds.), *Routes and Rites to the City: Mobility, Diversity and Religious Space in Johannesburg*. Palgrave Macmillan. London, pp. 137–162.

Johannesburg Development Agency, 2010. *Johannesburg Development Agency Annual Report 2010/11*. Johannesburg Development Agency. Newtown, Johannesburg.

Kelly, P., Lusis, T., 2006. Migration and the transnational habitus: evidence from Canada and the Philippines. *Environment and Planning A*. 38, 831–847.

Kenny, B., 2004. Selling selves: East Rand retail sector workers fragmented and reconfigured. *Journal of Southern African Studies*. 30, 477–498.

Kenny, B., Webster, E., 1998. Eroding the core: flexibility and the re-segmentation of the South African labour market. *Critical Sociology*. 24, 216–243.

Khan, F., 2000. *iGoli* 2002 – is the future private? *Green Left Online*.

Kipfer, S., 2008. How Lefevbre urbanized Gramsci: hegemony, everyday life, and difference, in: Goonewardena, K., Kipfer, S., Milgrom, R., Schmid, C. (Eds.), *Space, Difference, Everyday Life: Reading Henri Lefebvre*. Routledge. New York, pp. 193–211.

Kuljian, C., 2014. The Central Methodist Church, in: Harrison, P., Gotz, G., Todes, A., Wray, C. (Eds.), *Changing Space, Changing City: Johannesburg After Apartheid*. Wits University Press. Johannesburg, pp. 494–497.

Lees, L., 2012. The geography of gentrification: thinking through comparative urbanism. *Progress in Human Geography*. 36, 155–171.

Lees, L., Shin, H.B., López-Morales, E. (Eds.), 2015. *Global Gentrifications: Uneven Development and Displacement*. Policy Press. Bristol.

Lees, L., Shin, H.B., López-Morales, E., 2016. *Planetary Gentrification*. Polity Press. Cambridge.

Lefebvre, H., 1991. *The Production of Space*. Blackwell Publishers. Oxford.

Lefebvre, H., 2003. *The Urban Revolution*. University of Minnesota Press. Minnesota.

Leggett, T., 2003. *Rainbow Tenement: Crime and Policing in Inner Johannesburg (Monograph No. 78)*. Institute for Security Studies Criminal Justice Monitoring Project. Pretoria.

Lemanski, C.L., 2006. Desegregation and integration as linked or distinct? Evidence from a previously 'white' suburb in post-apartheid Cape Town. *International Journal of Urban and Regional Research*. 30, 564–586.

Lemanski, C.L., 2009. Augmented informality: South Africa's backyard dwellings as a by-product of formal housing policies. *Habitat International*. 33, 472–484.

Ley, D., 2003. Artists, aestheticisation and the field of gentrification. *Urban Studies*. 40, 2527–2544.

Lipietz, B., 2004. Muddling through: urban regeneration in inner-city Johannesburg. NAERUS Annual Conference. Barcelona.

Lipietz, B., 2008. Building a vision for the post-apartheid city: what role for participation in Johannesburg's City Development Strategy. *International Journal of Urban and Regional Research*. 32, 135–163.

Maharaj, B., Mpungose, J., 1994. The erosion of residential segregation in South Africa: the 'greying' of Albert Park in Durban. *Geoforum*. 25, 19–32.

Mamdani, M., 1996. *Citizen and Subject: Contemporary Africa and the Legacy of Late Colonialism*. Princeton University Press. Princeton, New Jersey.

Marom, N., 2014. Relating a city's history and geography with Bourdieu: one hundred years of spatial distinction in Tel Aviv. *International Journal of Urban and Regional Research*. 38, 1344–1362.

Matsipa, M., 2017. Woza! Sweetheart! On braiding epistemologies on Bree Street. *Thesis Eleven*. 141, 31–48.

Mayson, S.S., Charlton, S., 2015. Accommodation and tenuous livelihoods in Johannesburg's inner city: the 'rooms' and 'spaces' typologies. *Urban Forum*. 26, 343–372.

Mbembe, A., 2008. Aesthetics of superfluity, in: Nuttall, S., Mbembe, A. (Eds.), *Johannesburg: The Elusive Metropolis*. Duke University Press. Durham, North Carolina, pp. 37–67.

Mbembe, A., Nuttall, S., 2008. Introduction: Afropolis, in: Nuttall, S., Mbembe, A. (Eds.), *Johannesburg: The Elusive Metropolis*. Duke University Press. Durham, North Carolina, pp. 1–33.

McMichael, C., 2015. Urban pacification and 'blitzes' in contemporary Johannesburg. *Antipode*. 47, 1261–1278.

Merrifield, A., 2006. *Henri Lefebvre: A Critical Introduction*. Taylor & Francis. London.

Merrifield, A., 2013. The urban question under planetary urbanization. *International Journal of Urban and Regional Research*. 37, 909–922.

Miraftab, F., 2007. Governing post-apartheid spatiality: implementing City Improvement Districts in Cape Town. *Antipode*. 39, 602–626.

Mitchell, D., Heynen, N., 2009. The geography of survival and the right to the city: speculations on surveillance, legal innovation, and the criminalization of intervention. *Urban Geography*. 30, 611–632.

Mokonyama, M., Mubiwa, B., 2014. Transport in the shaping of space, in: Harrison, P., Gotz, G., Todes, A., Wray, C. (Eds.), *Changing Space, Changing City: Johannesburg After Apartheid*. Witwatersrand University Press. Johannesburg, pp. 194–214.

Morris, A., 1997. Physical decline in an inner-city neighbourhood. *Urban Forum*. 8, 153–175.

Morris, A., 1999a. *Bleakness & Light: Inner-City Transition in Hillbrow, Johannesburg*. Wits University Press. Johannesburg.

Morris, A., 1999b. Tenant–landlord relations, the anti-apartheid struggle and physical decline in Hillbrow, an inner-city neighbourhood in Johannesburg. *Urban Studies*. 36, 509–526.

Morris, A., 1999c. Race relations and racism in a racially diverse inner city neighbourhood: a case study of Hillbrow, Johannesburg. *Journal of Southern African Studies*. 25, 667–694.

Mosoetsa, S., 2011. *Eating from One Pot: The Dynamics of Survival in Poor South African Households*. Eurospan. Johannesburg.

Mosselson, A., 2017. 'Joburg has its own momentum': towards a vernacular theorisation of urban change. *Urban Studies*. 54, 1280–1296.

Murray, M.J., 2008. *Taming the Disorderly City: The Spatial Landscape of Johannesburg After Apartheid*. Cornell University Press. Ithaca, New York.

Murray, M.J., 2011. *City of Extremes: The Spatial Politics of Johannesburg*. Duke University Press. Durham, North Carolina.

Mushongera, D., 2017. Beyond GDP in assessing development in South Africa: the Gauteng City-Region Socio-Economic Barometer. *Development Southern Africa*. 34, 330–346.

Mushongera, D., Zikhali, P., Ngwenya, P., 2017. A Multidimensional Poverty Index for Gauteng Province, South Africa: evidence from Quality of Life Survey Data. *Social Indicators Research*. 130, 277–303.

Myers, G.A., 2011. *African Cities: Alternative Visions of Urban Theory and Practice*. Zed Books. London.

Naidoo, P., 2007. Struggles around the commodification of daily life in South Africa. *Revue of African Political Economy*. 34, 57–66.

Nevin, A., 2014. Instant mutuality: the development of Maboneng in inner-city Johannesburg. *Anthropology Southern Africa*. 37, 187–201.

Nuttall, S., 2004. City forms and writing the 'now' in South Africa. *Journal of Southern African Studies*. 30, 731–748.

Oldfield, S., Greyling, S., 2015. Waiting for the state: a politics of housing in South Africa. *Environment and Planning A*. 47, 1100–1112.

Paasche, T.F., Yarwood, R., Sidaway, J.D., 2014. Territorial tactics: the socio-spatial significance of private policing strategies in Cape Town. *Urban Studies*. 51, 1559–1575.

Parnell, S., Robinson, J., 2006. Development and urban policy: Johannesburg's city development strategy. *Urban Studies*. 43, 337–355.

Parnell, S., Robinson, J., 2012. (Re)theorizing cities from the Global South: looking beyond neoliberalism. *Urban Geography*. 33, 593–617.

Peck, J., 2015. Cities beyond compare? *Regional Studies*. 49, 160–182.

Peyroux, E., 2006. City Improvement Districts (CIDs) in Johannesburg: assessing the political and socio-spatial implications of private-led urban regeneration. *Trialog*. 89, 9–14.

Peyroux, E., 2012. Legitimating Business Improvement Districts in Johannesburg: a discursive perspective on urban regeneration and policy transfer. *European Urban and Regional Studies*. 19, 181–194.

Pieterse, E., 2011. Grasping the unknowable: coming to grips with African urbanisms. *Social Dynamics*. 37, 5–23.

Plissart, M.F., De Boeck, F., 2006. *Kinshasa: Tales of the Invisible City*. Ludion. New York.

Popke, J.E., Ballard, R., 2004. Dislocating modernity: identity, space and representations of street trade in Durban, South Africa. *Geoforum*. 35, 99–110.

Prabhala, A., 2008, Yeoville confidential, in: Nuttall, S., Mbembe, A. (Eds.), *Johannesburg: The Elusive Metropolis*. Duke University Press. Durham, North Carolina, pp. 307–316.

Prigge, W., 2008. Reading the urban revolution: space and representation, in: Goonewardena, K., Kipfer, S., Milgrom, R., Schmid, C. (Eds.), *Space, Difference, Everyday Life: Reading Henri Lefebvre*. Routledge. New York, pp. 46–61.

Propertuity, 2013. *The Maboneng Precinct Property Growth Report 2013*. Propertuity. Johannesburg.

Rérat, P., Lees, L., 2011. Spatial capital, gentrification and mobility: evidence from Swiss core cities. *Transactions of the Institute of British Geographers*. 36, 126–142.

Robinson, J., 1996. *The Power of Apartheid: State, Power, and Space in South African Cities*. Butterworth-Heinemann. Michigan.

Robinson, J., 2006. *Ordinary Cities: Between Modernity and Development*. Routledge. London.

Robinson, J., 2015. 'Arriving at' urban policies: the topological spaces of urban policy mobility. *International Journal of Urban and Regional Research*. 39, 831–834.

Robinson, J., 2016. Comparative urbanism: new geographies and cultures of theorizing the urban. *International Journal of Urban and Regional Research*. 40, 187–199.

Roy, A., 2009. The 21st-century metropolis: new geographies of theory. *Regional Studies*. 43, 819–830.

Rubin, M., Appelbaum, A., 2016. *Spatial Transformation through Transit-Oriented Development: Synthesis Report, Spatial Transformation through Transit-Oriented Development in Johannesburg Research Report Series. South African Research Chair in Spatial Analysis and City Planning*. University of the Witwatersrand. Johannesburg.

Savage, M., 2011. The lost urban sociology of Pierre Bourdieu, in: Bridge, G., Watson, S. (Eds.), *The New Blackwell Companion to the City*. Wiley Blackwell. Hoboken, New Jersey, pp. 511–520.

Schmid, C., 2008. Henri Lefebvre's theory of the production of space: towards a three-dimensional dialectic, in: Goonewardena, K., Kipfer, S., Milgrom, R., Schmid, C. (Eds.), *Space, Difference, Everyday Life: Reading Henri Lefebvre*. Routledge. New York, pp. 27–45.

Seekings, J., Nattrass, N., 2005. *Class, Race, and Inequality in South Africa*. Yale University Press. London.

SERI, 2013. *Minding the Gap: An Analysis of the Supply of and Demand for Low-Income Rental Accommodation in Inner City Johannesburg*. Socio-Economic Rights Institute of South Africa. Johannesburg.

Sihlongonyane, M.F., 2015. The rhetorical devices for marketing and branding Johannesburg as a city: a critical review. *Environment and Planning: Economy and Space*. 47, 2134–2152.

Simone, A., 2001. Straddling the divides: remaking associational life in the informal African city. *International Journal of Urban and Regional Research*. 25, 102–117.

Simone, A., 2004a. *For the City yet to Come: Changing African Life in Four Cities*. Duke University Press. Durham, North Carolina.

Simone, A., 2004b. People as infrastructure: intersecting fragments in Johannesburg. *Public Culture*. 16, 407–429.

Simone, A., 2008. The politics of the possible: making urban life in Phnom Penh. *Singapore Journal of Tropical Geography*. 29, 186–204.

Simone, A., 2010. A town on its knees? Economic experimentations with postcolonial urban politics in Africa and Southeast Asia. *Theory, Culture and Society*. 27, 130–154.

Smith, N., 2002. New globalism, new urbanism: gentrification as global urban strategy. *Antipode*. 34, 427–450.

Stanek, L., 2011. *Henri Lefebvre on Space: Architecture, Urban Research, and the Production of Theory*. University of Minnesota Press. Minnesota.

Steck, J.-F., Didier, S., Morange, M., Rubin, M., 2013. Informality, public space and urban governance: an approach through street trading (Abidjan, Cape Town, Johannesburg, Lome and Nairobi), in: Bekker, S., Fourchard, L. (Eds.), *Governing Cities in Africa: Politics and Policies*. HSRC Press. Cape Town, pp. 145–168.

Stoler, A.L., 2008. Imperial debris: reflections on ruins and ruination. *Cultural Anthropology*. 23, 191–219.

Teppo, A., Millstein, M., 2015. The place of gentrification in Cape Town, in: Lees, L., Shin, H.B., Lopes-Morales, E. (Eds.), *Global Gentrifications: Uneven Development and Displacement*. Policy Press. Bristol, pp. 419–440.

Tissington, K., 2009. *The Business of Survival: Informal Traders in Inner City Johannesburg*. Centre for Applied Legal Studies. Johannesburg.

Todes, A., 2014. The impact of policy and strategic spatial planning, in: Harrison, P., Gotz, G., Todes, A., Wray, C. (Eds.), *Changing Space, Changing City: Johannesburg After Apartheid*. Wits University Press. Johannesburg, pp. 83–100.

Tomlinson, R., Beauregard, R.A., Bremner, L., Mangcu, X., 2003. Introduction, in: Tomlinson, R., Beauregard, R.A., Bremner, L., Mangcu, X. (Eds.), *Emerging Johannesburg: Perspectives on the Postapartheid City*. Routledge. New York, pp. ix–xv.

van Holdt, K., 2010. The South African post-apartheid bureaucracy: inner workings, contradictory rationales and the developmental state, in: Edigheji, O. (Ed.), *Constructing a Democratic Developmental State in South Africa: Potentials and Challenges*. HSRC Press. Cape Town, pp. 241–260.

van Holdt, K., 2012a. Bodies of defiance, in: Burawoy, M., van Holdt, K. (Eds.), *Conversations with Bourdieu*. Wits University Press. Johannesburg, pp. 46–50.

van Holdt, K., 2012b. The Johannesburg moment, in: Burawoy, M., van Holdt, K. (Eds.), *Conversations with Bourdieu*. Wits University Press. Johannesburg, pp. 1–6.

van Holdt, K., 2013. The violence of order, orders of violence: between Fanon and Bourdieu. *Current Sociology*. 61, 112–131.

van Onselen, C., 2001. *New Babylon New Nineveh*. Jonathan Ball Publishers. Johannesburg.

Vicario, L., Martinez Monje, P.M., 2005. Another 'Guggenheim effect'? Central city projects and gentrification in Bilbao, in: Atkinson, R., Bridge, G. (Eds.), *Gentrification in Global Context: The New Urban Colonialism*. Routledge. London, pp. 151–167.

Viruly, F., Bertoldi, A., Booth, K., Gardner, D., Hague, K., 2010. *Analysis of the Impact of the JDA's Area-Based Regeneration Projects on Private Sector Investments: An Overview*. Johannesburg Development Agency. Johannesburg.

Walsh, S., 2013. We won't move: the suburbs take back the centre in urban Johannesburg. *City*. 17, 400–408.

Wilhelm-Solomon, M., 2016. Decoding dispossession: eviction and urban regeneration in Johannesburg's dark buildings. *Singapore Journal of Tropical Geography*. 37, 378–395.

Winkler, T., 2009. Prolonging the global age of gentrification: Johannesburg's regeneration policies. *Planning Theory*. 8, 362–381.

Zack, T., 2016. *Platform to an Arrival City: Johannesburg's Park Station and Surrounds, Spatial Transformation through Transit-Oriented Development in Johannesburg Research Report Series. South African Research Chair in Spatial Analysis and City Planning*. University of the Witwatersrand. Johannesburg

Zukin, S., 1998. Urban lifestyles: diversity and standardisation in spaces of consumption. *Urban Studies*. 35, 825–839.

2 An overburdened process

The competing agendas, imperatives and outcomes of inner-city regeneration

Introduction

During the recent tumultuous struggles over access to higher education in South Africa, 'the missing middle' emerged as a highly significant, politically charged category. It is a term used to describe students coming from households who earn too much to qualify for state bursaries and subsidised loans, but still cannot afford to pay for their university studies. It is an extremely broad category, including households earning between R120 000 and R399 996 annually.[1] Protesting students have consistently argued that students falling within this broad income bracket cannot afford university education, and have demanded that education should be universally free. In the face of these demands, some sections of government and university management have maintained the position that free university education for all is too costly, and that students coming from middle-class households must pay for their education, particularly as this will help subsidise poorer students. At the time I am writing (November 2017), these debates are still raging, with little sign of middle ground emerging.

I bring up this conflict in the context of a book about housing and urban development as it is illustrative of the competing demands placed on government, the expectations people hold in the post-apartheid period, the meagre resources which have to be stretched across numerous needs and contending factions, and the inequalities which define postcolonial society. These conflicts and competing demands are also prominent in the realm of housing, where the plight of a 'missing middle' is equally severe. Whilst the post-apartheid government has made extensive (although not always successful) efforts to provide housing for poor households (defined as earning less than R42 000 annually or R3500 per month), over 20% of South Africa's population fall within an income bracket which receives minimal government support, but is also largely excluded from the formal market. These households, commonly referred to as 'gap households' or the 'gap market,' generally earn monthly salaries between R3500 and R14 000 (i.e. R42 000–R168 000) (Centre for Affordable Housing Finance in Africa, 2015), and have become the key group which social and affordable housing companies involved in urban regeneration in the inner-city have targeted.

This chapter outlines how the redevelopment of the housing stock in the inner-city is being financed with this missing group in mind. It demonstrates that urban regeneration is being utilised as a vehicle for providing housing for lower-income groups, and therefore fulfils a powerful social and developmental mandate. However, it will also become clear that, despite the laudable intentions framing the process, regeneration and housing provision still occur largely within a market-based, commercial (broadly neoliberal) paradigm and do not cater to the poorest segments of society. As such, the financing behind urban regeneration and housing provision is shown to pursue contradictory, hard to reconcile practices. The housing being developed is centrally located, and thus offers a necessary alternative to the predominant forms of housing supported by government. It also plays a significant role in furthering spatial and racial integration, particularly on a city-wide level. Yet, when the process is situated in its local context, it comes to have deleterious effects on poorer communities living in the immediate area. Together, these competing agendas and practices make regeneration and housing provision an overburdened process which is trying to fulfil dissonant needs. It is therefore a process which cannot be summarised in a single meaning or result, but needs to be understood as an amalgamation of various vernacular impulses, logics and goals.

Housing after apartheid

Failing to reconcile multiple demands

Housing policy in the post-apartheid period has been formulated in response to three critical concerns: the need to provide adequate housing for millions of people who lack decent shelter, the urgent need to address racial and spatial segregation and transform South African cities, and the desire to compensate for historic injustices and provide homeownership opportunities to people who were discriminated against during apartheid (Huchzermeyer, 2001). Dealing with just one of these challenges is difficult enough, but trying to accommodate all three has proven a near-impossible task. Understandably, but not necessarily to the country's widespread benefit, government has focused on the first objective, prioritising the (relatively) swift construction of large numbers of owner-occupied houses which are distributed to the country's poorest households (Charlton, 2014; Charlton and Kihato, 2006).

The National Housing Subsidy Scheme (NHSS) forms the foundation of the ANC government's housing policy. In this project, private firms are subcontracted to construct stand-alone, four-room houses on state-owned land, which are then distributed to qualifying South African citizens. Colloquially referred to as RDP houses, these homes are available to South African citizens over the age of 21, living with dependents and earning less than R3500 per month. Since 1994 over 2 million NHSS houses have been constructed, to the benefit of almost 11 million people (Tissington, 2011), certainly a remarkable and significant achievement. However, following the contradictory nature of urban policies, the NHSS has been beset by a number of difficulties, many of which stem from the multiple agendas it is being used to fulfil. The process of acquiring an NHSS house is long and arduous, and has been characterised by inefficiencies and corruption. To be allocated a house, potential recipients are required to lodge applications at local municipal offices. They are then entered into a national housing database and allocated houses as and when they become available, which can often take decades. These long, indeterminate periods of waiting have become defining characteristics of citizenship in the post-apartheid period (Oldfield and Greyling, 2015), and frequently prompt communities to seek their own, informal solutions and vent their frustrations in violent public protests (van Holt et al., 2011). Delays in effectively providing state housing leave people in conditions of precariousness and contribute to the spread of informal settlements and shack houses constructed in people's backyards, as the country's growing population finds alternative ways to secure shelter. Thus, despite the relatively large number of units which have been provided, housing provision has not been able to keep pace with population increases. Currently, 13.6% of South Africa's population of 51 million continue to live in informal dwellings (Statistics South Africa, 2012).

In addition to delays in constructing and delivering housing, the NHSS has also failed to integrate beneficiaries into urban society. The programme remains caught between the tension of producing a large number of housing units quickly and efficiently, whilst simultaneously creating opportunities for upward social mobility and addressing the fragmentation and sprawl of South Africa's cities. NHSS housing projects are costly and absorb a great amount of government resources. They are, therefore, most frequently situated on marginal land, located far from urban centres, in some cases even beyond the peripheral townships (Tomlinson, 2006; Tomlinson et al., 2003). The areas in which NHSS housing settlements have been constructed are far removed from employment opportunities and lack social amenities such as schools, health-care facilities and recreational spaces. They have, then, maintained the marginalisation of the poor, who are disproportionally black, and have also enhanced the unequal, exclusionary geographies which define urban South Africa. It is a bitter paradox that one of the most important symbols of post-apartheid citizenship has, in practical terms, done little to enhance recipients' rights to the city (Huchzermeyer, 2001).

Regardless of the negative consequences engendered for a number of households, ideals of home ownership and suburban forms of housing remain deeply entrenched within the South African population's aspirations (Charlton, 2009). Thus, government is reluctant and even unable to move away from the four-room, stand-alone-unit typology. The competing rationalities of large-scale provision, satisfying people's expectations and aspirations, and using state housing to provide impetus for urban densification and inclusion continue to play out in the current debate surrounding 'mega' human settlements, the government's latest attempt to address housing shortages (Ballard and Rubin, 2017). The 'mega' human settlement strategy is a further iteration of the NHSS and is promoted as a way to speed up the construction of large quantities of houses. However, quantity of units and cost-efficiency (from the state's perspective), rather than locational quality, continue to be the defining principles. Hence, notwithstanding the body of evidence pointing out the inadequacies of the NHSS model, the proposed 'mega' human settlements are all located on peripheral land and urban edges.[2]

Yet, despite recent pronouncements favouring 'mega' human settlements, some policy debates have added nuance and come to emphasise the need for a variety of housing solutions in South Africa. In some sectors it has been recognised that the standardised approach of 'four-roomed houses' needs to be accompanied by access to rental accommodation, incremental upgrading of informal settlements and the development of low-income housing which is well connected and integrated into the rest of the urban environment (Tomlinson, 2006; Tissington, 2011). These additions were contained in the Breaking New Ground (BNG) strategy document, launched by the Department of Human Settlements in 2004. BNG introduced a new definition of and approach to housing. It places emphasis on the term 'human settlements' and highlights the fact that housing is multi-dimensional and has to provide for not only the forms of shelter people have access to, but the quality of their lives, their access to employment, amenities and resources and long-term environmental sustainability too (National Department of Housing, 2004).

The BNG strategy also included the goal of expanding the supply of rental housing (Tissington, 2011). Given emotive, aspirational issues around home-ownership, rental has not enjoyed popular support in the post-apartheid period. However, rental is a vitally important form of tenure, particularly in an economy in which few people have savings or income streams large and consistent enough for them to secure deposits on houses or maintain mortgage repayments. Furthermore, in a context of severe 'spatial mismatch,' where the majority of residential areas are far away from the bulk of industry and places of employment (SERI, 2016), rental is also important as it allows people flexibility and the possibility of relocating for work opportunities, whilst still maintaining homes in or connections with other areas. The development of a large stock of rental units is also important as it allows poorer households access to central regions of the city, where land and property prices are generally higher. This not only gives poorer households better access to amenities and employment opportunities, it also contributes to the growing push towards densification in South African cities. Recently local municipalities and national and provincial government have all emphasised the need for housing to be constructed in existing settled areas in their strategic planning (Huchzermeyer, 2014). Developing stocks of rental housing thus aligns with and compliments several important policies. However, these priorities are not necessarily shared by the broader population. Thus, government, torn between competing rationalities and needs, has remained fixed on the NHSS and 'mega' human settlement models, and support for rental housing has been allocated a fractional proportion of the Department of Human Settlement's budget.

The missing-middle: financing the gap market

There are other significant obstacles which make it difficult to provide the variety of housing typologies and solutions which are required. One of the biggest factors which constrains low-income households' opportunities to access decent housing is the shortage of loan and mortgage finance available to them (Centre for Affordable Housing Finance in Africa, 2015; Huchzermeyer, 2003; Smets, 2004). The NHSS rightly focuses on the most destitute and precarious members of South African society; but, as indicated in the Introduction, a large proportion of people are neglected by both the commercial market and government support programmes. The support that government is able to provide for the gap market comes in the form of a social housing subsidy which is allocated either on an individual household basis or to registered social housing institutions. Individuals are provided with grants worth between R10 000 and R87 000, which are calculated on a sliding scale, with beneficiaries who earn less being allocated more. The amounts available, though, are too small to cover the costs of purchasing a suitable house on the market and recipients have to secure additional finance from commercial banks. However, South African commercial banks, like their international counterparts (Smets, 1999), are reluctant to finance 'gap' households (Pillay and Naudé, 2006). Low-to-moderate-income households thus have few options and frequently resort to renting backyard shacks in townships, thereby exacerbating the sprawl of South Africa's cities and placing greater infrastructure burdens on already stressed areas, or downward raiding and either purchasing or renting NHSS houses from their original owners (Lemanski, 2014). Thus, whilst some progress has been made in broadening the scope of housing provision and seeking out alternative modalities, severe obstacles remain in place.

In addition to denying housing finance to low-income households, commercial banks also contributed directly to the process of urban decay in the inner-city. They adopted a policy of red-lining the area in the 1990s and continue to regard it as a dangerous investment destination (Murray, 2011). Low-income households are generally still unable to gain finance to acquire housing in the inner-city, and property developers too were prevented from gaining access to capital to fund the purchase and renovation of buildings. This created a chasm in the supply of housing for the low-income population and also led to the ongoing deterioration and destruction of the inner-city. Recognising the shortage of agencies willing and able to provide finance for gap housing, the South African government took the proactive step of establishing the National Housing Finance Corporation (NHFC) in 1996. This agency was originally founded with capital provided by government, with the purpose of providing loans and financial assistance to households who earned above the NHSS threshold but were unable to secure housing finance. They have also increasingly become involved in funding large-scale housing projects. Consequently, they are very active in the housing sector in Johannesburg's inner-city and helped finance the first private-sector-led acquisition and renovation of inner-city buildings. The agency is one of the main catalysts of urban renewal in Johannesburg and has played a fundamental role in shaping the direction and form of the process. They also continue to be one of the main sources of finance for social housing.

The NHFC do not act as the sole funders of housing projects, but rather guarantee loans and provide banks with assurance that, should a project fail, the portion they have financed will be paid back first. They therefore aim to attract finance back to areas like the inner-city and, in doing so, aid in expanding the supply of housing available to low-income earners. Based on the success of some NHFC-funded projects, some commercial banks are now willing to lend money for redevelopment in the area, provided projects are undertaken by large companies and are underwritten by the NHFC or other finance agencies. Whilst the NHFC is a national agency and operates throughout South Africa (although a very large proportion of their business has been concentrated in inner-city Johannesburg), two other finance agencies focusing specifically on the Johannesburg context have also been established.

The Trust for Urban Housing Finance (TUHF) has replaced the NHFC as the predominant source of finance for housing projects in the inner-city. The agency was born out of the Inner City Housing Upgrading Trust (ICHUT), an independent body which, unsuccessfully, sought to fund urban renewal and housing projects in inner-city Johannesburg in the late 1990s and early 2000s.[3] Learning from the negative experience of ICHUT, TUHF ensures that they have influence over the entire renovation process: they help potential clients develop business plans for the properties they want to purchase, give them training in running residential buildings and attempt to put developers in contact with commercial banks who can provide additional finance. Their holistic approach attempts to fulfil a number of agendas: helping nurture a new class of property owners and housing providers, entrenching commercial principles in the way housing is provided, and enhancing the banks' confidence in the inner-city, thereby creating more potential funding sources. Although they have supported large companies, they are also explicitly focused on racial redress and supporting emerging black entrepreneurs.

The Gauteng Partnership Fund (GPF) plays a similar role, although unlike TUHF, which is an independent commercial agency, it was established by the then Department of Housing (which was renamed the Department of Human Settlements in 2009) and is defined as a public institution. Like the NHFC and TUHF, they also seek to act as a bridge between the public and private sectors and leverage private finance for the construction of low-income housing. They provide 'nurse finance' which helps to initiate projects and cover early capital costs so that additional finance from the private sector can be secured. For example, the Johannesburg Housing Company's Brickfields development (a mixed-income housing project in Newtown, in the south-west of the inner-city) cost a total of R120 300 000, out of which R24 300 000 was provided by the GPF. With the various subsidy mechanisms in place it is now possible for up to 70% of the initial costs of social housing projects to be covered.

Additional subsidies for social housing institutions are provided in the form of the Capital Restructuring Grant, which is used to cover the initial costs of housing projects, particularly the purchase of land or property in strategic areas, such as the inner-city, or the upgrading of former single-sex hostels in townships (HDA, 2013; Tissington, 2011). The other nationally administered grant is the Institutional Subsidy (ibid.). In South Africa there is no continuous subsidy or housing benefit awarded to individual households; however, individual households can access a once-off subsidy grant or registered social housing institutions can apply for grants for individual projects which are aimed at households which fall within the specified income bracket. Again, this grant is not continuous, but serves to help cover initial capital costs. After these grants have been exhausted social housing institutions are expected to acquire additional funding from agencies such as the NHFC, TUHF and GPF and to become self-sustaining through viable commercial practices.

Conflicting regeneration agendas

Regeneration through the market

Thus, whilst there is a fair amount of state support for the social housing sector, significant emphasis is also placed on commercial principles and 'orthodox' market-based practices. As the CEO of TUHF explains, 'At its most basic, TUHF believes that if you address the causes of market failure – because there was substantial market failure in the inner-city – and if you introduce liquidity the market will work and property prices will increase.' They are therefore committed to the predominant pro-growth vision of regeneration developed by the City of Johannesburg. Criticising the way in which this approach has been adopted across South Africa's urban centres, the General Manager of the National Association of Social Housing Organisations (NASHO), a lobby group which works to promote the interests of social housing institutions, points out,

> Most of the strategic driving of inner-city rejuvenation is coming from municipalities and most of them have policies which say 'We will invest in infrastructure in order to increase the quality of the environment in order to get the private sector expenditure in,' and that investment doesn't include government-financed housing opportunities.

In Johannesburg specifically, the central components of local government's regeneration strategy are the Inner City Regeneration Charter (ICRC) and the more recent Inner City Roadmap (ICR). The overall emphasis in these plans is the need to attract private investment back into the inner-city. This is achieved by local government addressing the factors which detract from business' confidence, particularly crime and grime issues, and making investments in infrastructure which are designed to improve the area's image (City of Johannesburg, 2007). Following these principles, local government has undertaken several high-profile infrastructure projects in the inner-city. These include the creation of the Newtown cultural precinct, the Diamond and Fashion Districts in the eastern segments of the CBD and the flagship Mandela Bridge. These were all commissioned in the mid-2000s and were intended to show investors that government is focused on supporting the inner-city, and that private investments will be secure and profitable (Viruly et al., 2010). They are therefore strategies which align Johannesburg with neoliberal trends shaping other international cities, where, as Harvey (2006, 2012) argues, governments persist in investing public money into urban infrastructure so that private investors can reap profits.

New government-led investments have also been accompanied by several highly visible police campaigns aimed at by-law enforcement and rooting out petty criminality. Undocumented immigrants and informal traders have borne the brunt of these 'crackdowns' which are intended to send signals to private investors that the City is serious about safeguarding their interests and protecting the area from disruptive or harmful elements (Klaaren and Ramji, 2001; McMichael, 2015). Therefore, the process shares a great deal of similarities with experiences of urban renewal from other cities around the world. It resonates with the revanchist forms of policing documented in New York (Smith, 1996), Mumbai (Appadurai, 2000; Fernandes, 2004), Rotterdam (Schinkel and van den Berg, 2011), Quito and Guayaquil (Swanson, 2007) and São Paolo (Saborio, 2013), amongst others, and seems to bear out Smith's (2002) claim that the main goal of urban renewal projects in the neoliberal era is replacing poor populations in urban centres with money and investment. These features lead some critics to identify Johannesburg as falling within the sphere of global gentrification and to argue that the regeneration process underway is yet another iteration of the global spread of neoliberal urbanism (Winkler, 2009).

Conditional finance and attempts to create an inclusive inner-city

However, whilst sharing some of the similarities which define gentrification and revanchist renewal, the Johannesburg case departs from the dominant international narrative in marked ways. One of the main differences between regeneration in the inner-city and other well-documented cases is the way in which the rent-gap has been utilised. Studies of gentrification across a variety of contexts have found the rent-gap to be one of the decisive factors instigating reinvestment (Bernt, 2012; Lopez-Morales, 2011; MacLeod, 2002; Smith, 1987). In Johannesburg, whilst a substantial rent-gap or deflated property market existed in the 1990s and first decade of the 2000s, it was not sufficient to attract new investors into the area; the lack of finance available and the area's persistent stigma, instability and insecurity prevented redevelopment from taking place. Local government therefore had to stimulate interest and create demand for property. In doing so, whilst government was eager to see the values of properties increase, conditions were also attached to new housing developments, which, rather than pursuing gentrification, aimed to provide housing for lower-income communities.

In order to stimulate demand for property, Johannesburg's inner-city has been declared an Urban Development Zone (UDZ). Any tax-paying individual or entity whose property falls within the UDZ is able to claim reductions on taxable income that is generated from investments in the defined area (South African Revenue Service Legal and Policy Division, 2006). Whilst the promise of tax rebates and augmented profits is a conventional neoliberal mechanism designed to boost private investment, the UDZ scheme is also intended to target housing providers catering to the gap market and attempts to assist them in maintaining rentals at reduced rates. In addition, local government also launched a strategy which stimulated demand for derelict properties. Many of the buildings which are used at present to provide social housing were acquired through the Better Buildings Programme (BBP). In this programme, launched in 1997 and origin-ally known as the Bad Buildings Programme, the City Council took over derelict buildings whose arrears owed to the Council for services had become greater than their market value. These properties were then sold to pre-approved devel-opers for the difference between the arrears and the market price (Murray, 2008; Zack et al., 2009). The City Council also seized certain properties and sold them at auction to private developers. Sales were conditional on the purchasing companies signing agreements promising ongoing good management of the building and that rents would cater to communities qualifying for social housing after renovations were completed. The BBP thus augmented the subsidies available to social housing companies and enabled them to acquire properties at prices which allowed them to charge low rentals and still turn profits. The Johannesburg Housing Company (JHC) was able to acquire several buildings through this programme, which stands as an unconventional and creative use of the deflated property market.

The strategic use of derelict buildings also assists in spatial integration, which has become an increasing area of focus for national and local government, as well as finance agencies. Social housing policy in South Africa increasingly pays attention to the location of potential housing developments. Unlike the NHSS, in which the construction of a high volume of units is the biggest concern (Tomlinson, 2006; Ballard and Rubin, 2017), the goal of social housing is to provide accommodation which enables its beneficiaries to be integrated into the urban fabric. Hence, a representative from the NHFC points out that 'The major arguments [for social housing] have largely been around creating greater oppor-tunities for integration of our cities and helping people who would have been deprived of that access to have access to well-located economic and other opportunities.' Developing social housing and stimulating demand for inner-city properties thus contributes to realising densification and broader urban strategies aimed at integrating poorer communities.

Agencies providing finance for low-income housing are integral to this process. The Operations Manager at the NHFC explains that they focus on 'integrated housing solutions.' They thus move away from the NHSS model, which provides 'just a house,' and instead fund projects which will provide access to 'all the facilities, amenities, everything together.' Their mode of evaluating projects does not simply consider the profitability of the project: 'Our criteria for approving finance is, are all these things there? Is there transport? Is there work nearby, employment opportunities? What else is there in terms of facilities, hospitals, schools?' Because of this approach, Johannesburg's inner-city has been an extremely attractive location. Whilst it is still largely run-down, it still boasts better infrastructure than the peripheral township areas and is also the central transport hub for the city. Enabling low-income earners to gain accommodation in it helps to integrate them into urban society and thus promotes their right to the city and centrality (this point will be elaborated on in chapters 5 and 6). It is, therefore, an essential step in the restructuring of South African cities and achieving a form of post-apartheid redress. So, whilst market concerns are firmly entrenched in the way in which regeneration has been conceived and pursued in Johannesburg, there are also broader developmental goals at work which reflect the demands and imperatives of contemporary South African society.

In order to provide affordable, well-located housing and maintain rentals at low rates the NHFC, TUHF and the GPF provide loans which add only minimal costs and do not put pressure on developers to immediately realise profits. As there is no rent control in South Africa, finance organisations insist that projects qualifying for their assistance cannot charge initial rentals above R4500, which can then only be increased annually in line with average inflation rates. Companies securing loans are given lenient terms and extended periods to pay them back. Loans the NHFC makes to private developers are charged at the prime interest rate plus 2%, whilst social housing institutions are charged the prime interest rate with no additional costs. Private rental projects are also charged an administration fee amounting to 2% of the loan value, which is waived for social housing. The GPF grants recipients of their finance a three- to four-year moratorium on loan repayments to ensure that the projects are viable first. Thus, according to an employee of the GPF, the loans which are provided are concerned more with the creation of housing than return on investment or profitability.

Furthermore, housing finance agencies are also committed to urban regeneration which promotes social cohesion and has beneficial effects on local communities. They seek to inculcate sympathy for low-income households' precarious financial situations in their clients and will only support 'responsible' landlords who are committed to improving the inner-city and assisting their tenants. The CEO of TUHF maintains that what differentiates them from the commercial banks is that they take the social values and operating practices of their potential clients into consideration. When evaluating proposed projects, they do not only make assessments based on the feasibility and figures they are presented with, but also on the individuals behind the projects, their track records, interpersonal skills and general understanding of the housing environment in South Africa. As he explains,

> We have this saying, if a good landlord comes in with a bad deal we'll work with him to get it right; if a bad landlord comes in with a good deal we won't touch him with a 45-foot barge pole!

Because of the nature of the market that landlords funded through these entities deal with, being able to make judgements and consider the difficult circumstances of their tenants is important. So, according to TUHF,

> If somebody arrives and says 'I genuinely have a problem' [a good landlord] will say 'Fine, you can have a few weeks to pay me.' And it's being able to make those kinds of judgements, that interpersonal stuff [that makes or defines a good landlord].

The NHFC also adopts this sensitive, needs-based approach, leading to what one interviewee describes as 'a softening up of the code by which entrepreneurs are driven.'

It is therefore apparent that a form of developmental regeneration is being pursued, even though it is being done through the market. It can be read as a hybrid form of regeneration which fulfils diverse agendas and outcomes. One of the outcomes, as the quote above demonstrates, is the creation of a habitus which places emphasis on socially beneficial approaches to regeneration. Habitus is described by Bourdieu (1990, p. 66) as a 'feel for the game'; it is a way of understanding the codes of practice, mannerisms, forms of distinction and dispositions which are valued in a particular society, and inculcating these into one's everyday actions, personality and self-presentation. The possession and mobilisation of this type of knowledgeable habitus is what grants individuals cultural capital. Being a responsible, 'good' and socially aware landlord thus becomes a form of capital as it not only grants developers access to finance – i.e. economic capital – it also ensures that they conform to the demands and forms of distinction which are coming to prevail in the inner-city. As an interviewee describes it, the goal of investment and regeneration in the inner-city is 'to make sure there's housing provided and not to put people out on the street.' Demonstrating that this disposition has been internalised, one private developer states, 'I think we work towards a common goal which is more, yes, the profit is necessary – at the end of the day everyone wants profit – but it's more of [sic] rejuvenating the inner-city.' The code of practice that has emerged is also enforced by finance agencies, as if developers fail to buy into the agenda, their reputations will be damaged, and this loss of cultural capital will also result in a loss of economic capital. In these ways, finance agencies are determining the fields in which housing providers and developers act, and the terms on which they are required to do so.

Responding to spatial and social realities

The devalued property market in the inner-city therefore has not been used in the way which has become commonplace around the world, including in other South African cities. Cape Town, for example, stands as an example of urban regeneration based predominantly on market concerns and the maximisation of property values and profits. This has created a situation in which low-income populations are excluded from the central city, both proactively through intensive policing, as well as through ever-increasing property values and ground rents (Miraftab, 2007; Paasche et al., 2014; Teppo and Millstein, 2015). In Johannesburg the experience has been different. In part this is due to the fact that the area underwent a process of transition which was far more destructive and violent than that experienced in Cape Town. During the period of transition from apartheid to democracy, Johannesburg's white residents and business owners were more inclined to abandon the area and relocate to the burgeoning suburbs springing up all around the wider metropolitan area (Morris, 1999). The relocation of the major enterprises and the Stock Exchange was also made by possible by the creation of a new business district in Sandton, which has attracted far more investment and property speculation (Goga, 2003). Cape Town, on the other hand, has been far more successful in maintaining the central city as the focal point of the economy. Since 2009 the city has also been run by the Democratic Alliance. This party is more unashamedly socially and economically (neo)liberal than the ANC, who have been the party in power in Johannesburg since 1994. Consequently, businesses and property owners have been given influential roles to play in running Cape Town's central city, as illustrated by the predominance of the Cape Town Partnership in driving regeneration efforts (Morange and Didier, 2006). Their programmes have been successful from a business perspective (although harmful in social terms) and, together with the area's natural beauty, have established the city as one of the most popular tourist destinations in the world.

In contrast, Johannesburg's inner-city was overrun with crime, drugs and violence and today the area continues to be saddled with a notorious reputation for being a den of vice, violence and social problems which repels many investors and visitors. This stigma has come to define the inner-city in many people's imaginations and imbue the place with a particular socio-cultural meaning. Whilst this contributes to the area's decay, it has also come to serve a socially beneficial function of sorts, as it has ensured that demand for accommodation in the inner-city, especially in the most densely populated, yet also most notorious, suburbs of Hillbrow and Berea only comes from low-income earners who need to live in the area out of economic necessity. Whilst central Cape Town has been maintained as an exclusive area and has almost no low-income housing, inner-city areas like Yeoville, Hillbrow and Berea have thus far failed to attract high-end developments and upper-class residents. Rather, the demand for accommodation in these areas comes from low-income earners, and finance agencies and housing providers have taken cognisance of this. They recognise that high-end redevelopments are not feasible and simultaneously actively discourage them in favour of more socially beneficial forms of regeneration. They are therefore responding to the realities of the market.

These market and social–spatial realities have proved persuasive, even to powerful, ideologically motivated actors. In August 2016 there was a change in government, with the ANC losing control of the City of Johannesburg for the first time in the post-apartheid period. A surprising coalition between the left-wing Economic Freedom Fighters and centre-right Democratic Alliance allowed

Herman Mashaba, a wealthy black entrepreneur, to become mayor of the city. In his short tenure so far, Mashaba has sought to move the inner-city regeneration agenda forward. He has made several public pronouncements on the issue and made it one of the key priorities of his administration. He initially offered rather vague, ambitious plans, admittedly based on the DA's experiences of governing Cape Town.[4] These included turning the inner-city into an international tourist attraction and shopping hub. As the realities of governing have become more apparent, the stance being offered by the current administration has softened and become more attuned to the functions the space is currently serving and is best-placed to continue to serve. Although the current policy keeps fealty with the direction established by the ANC-led administrations and focuses on enhancing the commercial viability of the area, cognisance that the inner-city is not going to attract wealthy residents has filtered through and the importance of the area for housing lower-income families has been entrenched in policy and planning approaches. Thus far, the administration's focus has been on by-law enforcement and transferring state-owned properties to affordable housing companies for redevelopment. On a grander scale, the mayor has vowed to turn the inner-city 'into a building site,'[5] but one in which housing opportunities for lower-income households are prioritised.

It therefore needs to be recognised that market realities and governing practices are being determined by the spatial conditions of the inner-city. A central ambition of Bourdieu's theoretical project was to establish a social understanding of the economy (Bourdieu, 2005; Reed-Danahay, 2005). The case of the inner-city demonstrates that in addition to prevailing socio-cultural stratifications, forms of habitus and capital, spatial conditions also need to be factored into understandings of how markets and demand work and economic agents function. Space thus needs to be regarded as a central component which shapes socio-cultural expectations, definitions and habitus. The practices of finance agencies show how the market is actually subject to and produced by spatial constraints, which actors have to adapt to. Finance agencies eschew high-end regeneration as they recognise it is not economically feasible. Whilst doing so, they also cultivate practices which entrench this recognition in housing providers and create an alternative framework for regeneration. They are able to take advantage of the rent-gap and low demand from wealthy people to expand the supply of social and affordable housing. Because the prices of property in the area are low, housing providers can charge lower rentals; the low costs of finance and relaxed repayment conditions also ensure that rentals can be maintained at reduced rates. Their approaches to providing housing are therefore adaptations to the prevailing material and social conditions of the inner-city; as these adaptations become established as standard practice and valorised as the correct types of behaviour, forms of habitus and cultural capital which reflect more developmentally focused goals emerge. This example furthers the argument that gentrification, which is frequently portrayed as a function of the capitalist market, rent-gap and an inexorable global process, is actually highly context-dependant and occurs only when political and spatial conditions are conducive to it (Bernt, 2012; Maloutas, 2012). The role space plays in structuring, permitting, restricting and influencing socio-cultural and economic practices therefore needs to be foregrounded in explanations of gentrification and regeneration.

Competing concerns and limitations to the developmental potential of regeneration

Between transformation and the market

Concerted efforts have been made to ensure that developmental regeneration takes place and some of the city's enduring spatial divisions are reduced. But there are also considerable tensions within the regeneration process and the ways in which it is unfolding. Whilst an interventionist approach has been adopted and investors and housing developers are not presented with blank cheques, commercial logics still prevail. South Africa lacks a standard or coherent policy on urban regeneration (HDA, 2013); rather, regeneration goals and plans are included in various city-wide documents and policies. Certainly, urban regeneration cannot take place in isolation from other planning and development procedures (Turok and Robson, 2007), but the lack of a focused regeneration strategy means that frequently there is too little attention paid to the renewal of central urban areas and the strategic possibilities these contain.

For instance, the General Manager of NASHO argues that, in contrast to the prevailing goal of increasing property values in the inner-city, a more focused policy of utilising state-owned land or buildings in the inner-cities for social housing is required. In contrast to the current situation, in which public investments in infrastructure are made with the aim of attracting private redevelopment, he proposes that a certain proportion of buildings in urban areas should be reserved for social housing and that those upgrading projects will then take the lead in stabilising inner-city areas and attracting other forms of investment. The course being pursued at present, whilst emphasising the need for affordable housing to be developed, does not make any specific commitments or strategic decisions around reserving buildings for social housing, and instead relies on the private sector to set rentals at appropriate levels. As the following chapters will show, and as the interviewee quoted here argues, social housing has been a very successful driver of urban renewal and improving community cohesion in the inner-city, and therefore does hold significant potential to be a catalyst for wider regeneration. However, the improvements that have come about have occurred without any specific planning frameworks in place to ensure that the successes achieved are replicated and the building stock which remains undeveloped is preserved for these purposes. Instead, a situation currently prevails where the commercial imperatives of the market, whilst restrained to some degree, are left intact. Thus, TUHF's CEO explains that whilst the agency is dedicated to providing low-income housing, they continue to operate within a strict financial paradigm. He explains the agency's outlook as follows:

TUHF is one of those organisations that we're a hard-nosed commercial organisation but we're also an organisation that worries a lot about doing good. So we have a very strong development objective in addition to our commercial objectives. That's why somewhere here you'll see the 'Good Business Doing Good' logo. So the 'Good Business' is hard-nosed commercial, growing, profitable, all those good things, and the 'Doing Good' is low-income housing, urban regeneration, access to finance.

They therefore seek to achieve divergent goals through regeneration, which places them in a contradictory position. For example, communal housing – where tenants rent bedrooms and share common bathroom and kitchen facilities – is the cheapest formal accommodation available in the inner-city. Rents in these buildings, which are provided by social housing and private companies, range between approximately R800 and R1700 (Tissington, 2011). However, TUHF does not finance communal developments as it fears that if the projects fail the resale value of these buildings will be too low. In this case, commercial concerns trump developmental imperatives. Furthermore, because they cater to a 'risky' market, the finance TUHF provides actually comes to be more expensive over time than that provided by commercial banks. The NHFC too, whilst being a government-aligned entity, has to be self-sustaining. Its operations are funded by the returns they make on the loans and investments they provide. Thus, they have to ensure that any project they back is profitable over the long term and they are prevented from financing housing which caters to the poorest members of society. Consequently, outside of the informal rental market, there are few options available to people earning below R3500 per month, even though they make up a substantial proportion of the inner-city's population.

Additionally, demand for accommodation and competition for space in the inner-city are such that there are very few vacant buildings. Thus, whilst the stated goal is not to put people out on the street, in reality, almost every commercial and social housing redevelopment has been preceded by an eviction. In some cases, evictees have been able to secure accommodation in other formalised buildings and have even been offered preferential placement and relocation assistance by housing companies. In other cases, however, people have literally been thrown out, together with all their meagre belongings and been forced to leave the area or seek shelter in other precarious, dilapidated buildings (COHRE, 2005). Thus, even though the process has been formulated with developmental concerns in mind, it has not failed to prevent displacement. Finance professionals consequently find themselves caught between the competing imperatives which shape the field in which regeneration is taking place.

Employees of both the GPF and NHFC speak of the need to start recouping the funding given to social housing institutions, as the practice of giving them loans which are not market-related is unsustainable over the long term. They note that government cannot afford to continue to subsidise social housing indefinitely and that it needs to be a commercially viable sector. The rationality or habitus they act within is clearly market-oriented, as is the dominant rationality shaping a large proportion of the regeneration strategies being pursued. On the other hand, they also have social concerns and are influenced by a form of habitus which is committed to development. As the NHFC's General Manager points out, 'I come from the perspective of saying my satisfaction is seeing people who previously didn't have housing having housing.' But, at the same time, the NHFC and other finance institutions 'cannot get involved in something that will lose a lot of money.' They therefore have to be cautious and put business concerns before social needs. As governments' fiscal constraints have become more acute in recent years, the business aspect of their work has taken greater precedence. State support for the NHFC has been reduced and they have subsequently been forced to borrow a greater proportion of money from commercial sources. Unlike government funding, this money is not interest-free and places pressure on them to be more commercially focused and consistent in generating profits. They are thus in a contradictory position and their ability to finance projects which may be risky or offer less return on investment is reduced. The aforementioned inter-viewee reflects on this predicament and soberly concludes 'there is a *dissonance* between what we're trying to do on the social side, on the one hand, and what we need to do on the other hand to keep ourselves sustainable.'

Reliance on market mechanisms to drive renewal also imposes some limita-tions on the extent to which the process can be maintained and expanded. The rent-gap and government support initially attracted investors to the area and made the provision of low-income housing feasible. However, as the inner-city has stabilised and more investors have entered the market, competition for buildings has increased and prices have gone up. An employee of the predomi-nant social housing company in the inner-city expresses frustration with the situation and notes that the increased competition for buildings and escalating property prices are hindering their ability to create more social housing in the area. He states, 'We don't raise rental so [increased costs and property prices] are killing it [social housing] at the feasibility stage.' Private housing providers who do not qualify for subsidies from the government also experience challenges to their business models as a result of increased prices. A small proportion acquired properties through the BBP, but the majority were purchased on the open market. When the bulk of the buildings currently used for low-income housing were purchased 'there were a lot of hijacked buildings, buildings that were valueless, worthless in market terms.' Developers were consequently able to acquire buildings for extremely low costs, which effectively acted as a type of subsidy allowing them to charge lower rentals and still generate profits. But as regenera-tion has taken hold and the area has become more attractive to investors, initial expenses and operating costs have risen, leading to rentals also increasing to cover these expenses. As a result, low-income housing becomes less feasible and harder to replicate in buildings which are yet to be renovated. Increasingly, then, the most significant challenge in the inner-city, as NASHO's General Manager explains, is finding ways to 'create the mechanisms that allow you to deal with a private property market that's getting better and better but at the same time protect the rights of people on lower incomes to access the city.' These instruments are limited at present and are largely confined to the conditional finance discussed in this chapter. Despite the presence of other ambitious policy imperatives, local government's focus remains on implementing orthodox market practices. Again, there is a tension between the potential transformative benefits of regenerating the inner-city through providing low-income housing and opening the area up to the private property market.

Commitment to economic orthodoxy is also apparent in the prevailing view that the land market in the inner-city should resemble market values and geographic patterns in the rest of the world. This means, as the Johannesburg Housing Company's (JOSHCO) CEO explains, 'It's a question of the value of space: access to the inner-city comes at a price, and the price of space in the Johannesburg inner-city must be greater [than in outlying areas].' There are thus two conflicting visions and programmes at play simultaneously – one is the powerful, democratic vision of an integrated city which includes low-income households in the most central areas and thus integrates them into urban society and reduces some of the spatial inequalities which define the city; the other is a strong acceptance of and commitment to a spatial landscape in which the central city is economically vibrant and valuable and the poor are confined to the geographic (and thus social) margins. This is the pattern which apartheid was built on and is still being maintained through contemporary approaches and policies.

Tensions between regenerating the inner-city and addressing city-wide concerns

It is apparent, then, that the regeneration process is overburdened in terms of the expectations and functions it is intended to fulfil. In one instance it represents a meaningful attempt to introduce an alternative means of housing provision into the city's landscape and enhance the abilities of lower-income households to access centrally located, well-resourced housing. At the same time, however, it is also a process which operates on a commercial basis and is intended to boost the value of inner-city property. The scope for accommodating poorer segments of the population is consequently reduced, and some of the progressive gains which have been made are in danger of being reversed. Regeneration is being framed in normative terms which regard the inner-city as too valuable to accommodate low-income residents. Given the realities of the area and the needs of the population who inhabit it, these ambitions are inappropriate and are exacerbating social problems and forms of exclusion. However, when the process is viewed in wider geographic and economic perspective, its logic has some justification.

As the inner-city starts to flourish and become dominated by private businesses, which are increasingly able to take care of maintenance issues themselves by contracting private cleaning, maintenance and security services in the CID areas, local government is freed up to concentrate resources and efforts on other neglected areas, particularly peripheral townships and informal settlements. The South African government is burdened with incredible socio-economic problems and also has severe budgetary constraints. Consequently, local government is increasingly looking to use the inner-city property market to fund operations elsewhere. The view prevails that, because they are now attractive locations for private investment, 'the inner-cities, in one way or another, will look after themselves and will look after themselves better than squatter camps [i.e. informal settlements],' as an advisor to the Minister of Human Settlements explained.

Rising property rates and value extracted from the inner-city are being regarded as tools for realising a progressive form of cross-subsidisation. During apartheid, municipal structures ring-fenced wealthy white suburbs and ensured that rates contributions were spent in the areas in which they were collected (Tomlinson et al., 2003). The restructuring of Johannesburg's municipalities to undo this deliberately unequal system and create a larger unified metropolitan system is one of the major achievements of post-apartheid urban governance (Beall et al., 2002; Lipietz, 2008). The metro system relies heavily on the rates and taxes charged on properties (which currently account for 18.5% of the CoJ's revenue stream and operating budget (City of Johannesburg, 2017)[6]). Thus, as property values in the inner-city rise, local government is able to increase the levies imposed on owners and augment its income.[7] For example, one medium-sized for-profit housing company pays the City of Johannesburg R2 million a month; the larger ones would be paying even more. Therefore, there is a great deal of benefit which can be gained from the rising property values in the inner-city, as they create more income for local government, which can potentially be used to address more immediate needs, including maintenance, service provision and infrastructure and informal settlement upgrading. Local government therefore has a direct interest in seeing property rates in the inner-city increase and the private sector flourish. The formalisation of housing in the inner-city also allows for more efficient and regular billing and payment for water, electricity and waste management services. At present, most residential buildings are billed directly for water and electricity; these charges are added onto tenants' monthly rentals and the City is able to efficiently collect on the rates owing.[8] This system is much more controlled and regularised than in other areas of the city, especially Soweto, which has a long history of rent and rate boycotts and where residents continue to resist paying for municipal services.[9] At present, formal buildings in the inner-city are subsidising other more dysfunctional, disputed and poor parts of the city.

In a context of severe spatial inequalities and a formerly fragmented and discriminatory approach to local governance, where wealthy white areas enjoyed far higher standards of service and maintenance, cross-subsidisation within the municipal region is a progressive practice and helps redistribute resources towards poorer areas. However, this strategy is failing to account for the fact that the majority of the population residing in the inner-city also survive on restricted incomes. They are coming under increased financial strain as ever-rising service charges are passed on to them. This affects the affordability of inner-city accommodation and further threatens to erode the gains made in providing low-income housing in the area. The overburdened, contradictory nature of regeneration is therefore apparent, as are the deleterious consequences. This is summed up bitterly but succinctly by the CEO of TUHF, who gives a powerful indictment of the limitations of the process, which he himself is actively reproducing. He declares

> [displacement has] been huge, it's inevitable, it's consistent with city strategy. You know, people wring their hands about the very poor, 'Create housing for them,' etc. etc. but actually, nobody gives a damn! What they like to see is that the city continues to regenerate, rents and property prices continue to go up, rates and taxes continue to go up, which feeds local government coffers.

In this way, he gives a clear indication of the conflicted nature of the regeneration process and the way it reflects and reproduces a dualistic social order which is caught between the logics of the market and the developmental needs and ambitions of the post-apartheid city.

Conclusion: neither the market nor development prevail

Thus, it can be concluded that neither the neoliberal, market-based approach nor the developmental agenda prevail. Rather, they simultaneously overlap and compete with one another, making the process an uncomfortable fit with pre-given or standardised definitions or descriptions. Powerful neoliberal, market-based influences, forms of knowledge and habitus certainly are playing significant roles in directing the process. At the same time, however, these influences do not occlude possibilities for developmental and transformative agendas to be pursued and for these goals to be realised. As Colomb (2007, p. 5) points out, the agendas behind urban regeneration strategies are frequently 'potentially complementary, but also contradictory.' This is certainly the case in inner-city Johannesburg, where, due to the finance and policy regimes which frame it, regeneration is a hybrid, complex process which reflects competing demands and needs. On the one hand, finance for regeneration is granting low-income people access to stable, formal and decent accommodation in the central city, which is a highly commendable and significant achievement. On the other, regeneration is being used to enhance the value of property in the area, attract more investment and increase tax income for the City Council; these measures are vital in the wider context of the city, but are putting financial strain on the people who are living in the immediate area and leading to displacement. It is therefore an overburdened process which has to accommodate many competing needs and which gives rise to contradictory outcomes.

These competing needs reflect the particular dynamics of the setting and social context in which the process occurs. They demonstrate that features and logics which reflect global processes and experiences, such as neoliberal forms of urban governance, property-based regeneration strategies and efforts to capitalise on a rent-gap, are always localised and grounded in the particular conditions and spaces in which they take place. They are therefore absorbed and translated into vernacular approaches and actions. Thus, the ways in which urban processes unfold and the outcomes they engender cannot be seen as fitting into pre-determined frameworks. The urban regeneration strategies, approaches to utilising the deflated property market and financing the provision of low-income housing in Johannesburg demonstrate that gentrification is not simply an inexorable global phenomenon and inevitable outcome of urban renewal processes, even ones in which strategies focusing on private investment and property-led redevelopment prevail. Rather, when gentrification does come about it needs to be understood as the product of deliberate social and political actions and decisions (Bernt, 2012; Maloutas, 2012). The case of Johannesburg shows that alternative decisions and agendas are not only possible, but already exist, and emphasises the need for critiques to be aware of the competing rationalities and agendas which are always present in urban processes and policies.

The following chapter will explore these alternative agendas and strategies in more detail and demonstrate how housing providers, from both the social and for-profit sectors, have responded to the conditions imposed by finance agencies. It shows how the competing agendas which are framing regeneration, and which circulate in the post-apartheid context, come to shape their dispositions, aspirations and actions – in other words, their habitus and praxis. Like the employees of finance agencies discussed in this chapter, housing providers are shown to operate in ways which reflect and reproduce commercial logics and market-based approaches to housing provision. At the same time, they also demonstrate commitment to the developmental goals of creating an inclusive city and providing decent, affordable housing. They are therefore shown to have a split, contradictory habitus, which is remaking the inner-city accordingly.

Notes

1 www.news24.com/PartnerContent/are-you-part-of-the-missing-middle-20160715.
2 For a detailed mapping of the proposed locations of future mega-human settlements in Gauteng see www.gcro.ac.za/outputs/map-of-the-month/detail/the-location-of-planned-mega-housing-projects-in-context/.
3 ICHUT provided the bridging finance for the Seven Buildings Project, a programme which purchased and renovated seven inner-city buildings and established them as housing cooperatives (Oelofse, 2003). However, the pilot project eventually collapsed as tenants refused to accept rental increases, which made ongoing maintenance impossible, and loan repayments to ICHUT ceased. All seven buildings became mired in in-fighting and accusations of theft and corruption amongst the cooperatives' management bodies and eventually went into liquidation (Oelofse, 2003). According to interviewees, this experience demonstrated the need for a more focused and systematic approach to financing, which would avoid making bad or risky loans, but would still facilitate the provision of low-income housing.
4 www.dailymaverick.co.za/opinionista/2016-05-08-reimaging-johannesburg-as-a-tourist-destination/.
5 www.dailymaverick.co.za/opinionista/2017-09-27-great-expectations-for-the-jozi-inner-city/#.WlOGXFVl_Gg.

6 The remaining bulk of the City's revenue comes from the charges attached to water, electricity and refuse collection services, which collectively make up 58.8% of the City's overall budget.

7 The City's reliance on rates as a source of revenue is illustrated by the effects of a boycott in the mid-1990s. In opposition to plans to unify the city's municipal areas and bring them all under one central body, residents in wealthy suburbs withheld their rates payments. Local government's income streams were so severely damaged as a result that the City was almost reduced to bankruptcy and could only begin to function once the boycott ended and rates started flowing again (Tomlinson et al., 2003).

8 Some housing companies have started installing pre-paid electricity meters in their apartments. Whilst these devices are criticised for the ways in which they enforce commodification and punitive user-pays principles (Hart, 2002), they are also increasingly in demand due to the crisis in the City's billing department. For several years the City has been unable to provide reliable billing for electricity and water, and instead issues irregular bills which arrive haphazardly, fluctuate wildly and seldom reflect the actual amounts people have used. For these reasons, pre-paid meters have actually become desirable; many tenants interviewed in this study called for pre-paid meters to be installed in their flats so that they could more easily budget and manage their consumption.

9 A recent news article estimated that residents in Soweto collectively owe the national electricity provider, Eskom, approximately R3.6 billion in unpaid bills. See http://mg.co.za/article/2014-11-19-sowetos-unpaid-bills-add-to- eskoms-financial-woes/.

References

Appadurai, A., 2000. Spectral housing and urban cleansing: notes on millennial Mumbai. *Public Culture.* 12, 627–651.

Ballard, R., Rubin, M., 2017. A 'Marshall Plan' for human settlements: how megaprojects became South Africa's housing policy. *Transformation: Critical Perspectives on Southern Africa.* 95, 1–31.

Beall, J., Crankshaw, O., Parnell, S., 2002. *Uniting a Divided City: Governance and Social Exclusion in Johannesburg.* Earthscan. London.

Bernt, M., 2012. The 'double movements' of neighbourhood change: gentrification and public policy in Harlem and Prenzlauer Berg. *Urban Studies.* 49, 3045–3062.

Bourdieu, P., 1990. *The Logic of Practice.* Stanford University Press. Stanford, California.

Bourdieu, P., 2005. *The Social Structures of the Economy.* Polity Press. Cambridge.

Centre for Affordable Housing Finance in Africa, 2015. *Understanding the Challenges in South Africa's Gap Housing Market and Opportunities for the RDP Resale Market.* Centre for Affordable Housing Finance in Africa. Johannesburg.

Charlton, S., 2009. Housing for the nation, the city and the household: competing rationalities as a constraint to reform? *Development Southern Africa.* 26, 301–315.

Charlton, S., 2014. Public housing in Johannesburg, in: Harrison, P., Gotz, G., Todes, A., Wray, C. (Eds.), *Changing Space, Changing City: Johannesburg After Apartheid.* Wits University Press. Johannesburg, pp. 176–193.

Charlton, S., Kihato, C., 2006. Reaching the poor? An analysis of the influences on the evolution of South Africa's housing programme, in: Pillay, U., Tomlinson, R., du Toit, J. (Eds.), *Democracy and Delivery: Urban Policy in South Africa.* HSRC Press. Cape Town, pp. 252–282.

City of Johannesburg, 2007. *Inner City Regeneration Charter.* City Council. Johannesburg.

City of Johannesburg, 2017. *City of Johannesburg Medium Term Budget 2017/18–2019/20.* City Council. Johannesburg.

COHRE, 2005. *Any Room for the Poor? Forced Evictions in Johannesburg, South Africa.* Centre for Housing Rights and Evictions. Johannesburg.

Colomb, C., 2007. Unpacking new labour's 'Urban Renaissance' agenda: towards a socially sustainable reurbanization of British cities? *Planning Practice and Research.* 22, 1–24.

Fernandes, L., 2004. The politics of forgetting: class politics, state power and the restructuring of urban space in India. *Urban Studies.* 41, 2415–2430.

Goga, S., 2003. Property investors and decentralization: a case of false competition? in: Tomlinson, R., Beauregard, R.A., Bremner, L., Mangcu, X. (Eds.), *Emerging Johannesburg: Perspectives on the Postapartheid City.* Routledge. New York, pp. 71–84.

HDA, 2013. *Reviving Our Inner Cities: Social Housing and Urban Regeneration in South Africa.* Housing Development Agency. Johannesburg, South Africa.

Hart, G., 2002. *Disabling Globilization: Places of Power in Post-Apartheid South Africa.* University of California Press. Berkley and Los Angeles, California.

Harvey, D., 2006. *The Limits to Capital.* Verso. London.

Harvey, D., 2012. *Rebel Cities: From the Right to the City to the Urban Revolution.* Verso. London.

Huchzermeyer, M., 2001. Housing for the poor? Negotiated housing policy in South Africa. *Habitat International.* 25, 303–331.

Huchzermeyer, M., 2003. Addressing segregation through housing policy and finance, in: Harrison, P., Huchzermeyer, M., Mayekiso, M. (Eds.), *Confronting Fragmentation: Housing and Urban Development in a Democratic Society.* University of Cape Town Press. Cape Town, pp. 211–227.

Huchzermeyer, M., 2014. Changing housing policy in South Africa, in: Bredenoord, J., Van Lindert, P., Smets, P. (Eds.), *Affordable Housing in the Urban Global South: Seeking Sustainable Solutions.* Routledge. London, pp. 336–348.

Klaaren, J., Ramji, J., 2001. Inside illegality: migration policing in South Africa after apartheid. *Africa Today.* 48, 34–47.

Lemanski, C., 2014. Hybrid gentrification in South Africa: theorising across southern and northern cities. *Urban Studies.* 51(14), 2943–2960.

Lipietz, B., 2008. Building a vision for the post-apartheid city: what role for participation in Johannesburg's city development strategy. *International Journal of Urban and Regional Research.* 32, 135–163.

Lopez-Morales, E., 2011. Gentrification by ground rent dispossession: the shadows cast by large-scale urban renewal in Santiago de Chile. *International Journal of Urban and Regional Research.* 35, 330–357.

MacLeod, G., 2002. From urban entrepreneurialism to a 'revanchist city'? On the spatial injustices of Glasgow's renaissance. *Antipode.* 34, 602–624.

Maloutas, T., 2012. Contextual diversity in gentrification research. *Critical Sociology.* 38, 33–48.

McMichael, C., 2015. Urban pacification and 'blitzes' in contemporary Johannesburg. *Antipode.* 47, 1261–1278.

Miraftab, F., 2007. Governing post-apartheid spatiality: implementing City Improvement Districts in Cape Town. *Antipode.* 39, 602–626.

Morange, M., Didier, S., 2006. Security discourses, community participation and the power structure in Cape Town, 2000–2006. *Urban Forum.* 17, 353–379.

Morris, A., 1999. *Bleakness & Light: Inner-City Transition in Hillbrow, Johannesburg.* Wits University Press. Johannesburg.

Murray, M.J., 2008. *Taming the Disorderly City: The Spatial Landscape of Johannesburg After Apartheid.* Cornell University Press. Ithaca, New York.

Murray, M.J., 2011. *City of Extremes: The Spatial Politics of Johannesburg*. Duke University Press. Durham, North Carolina.

National Department of Housing, 2004. 'Breaking New Ground.' A Comprehensive Plan for the Development of Sustainable Human Settlements.

Oelofse, M., 2003. Social justice, social integration and the compact city: lessons from the inner city of Johannesburg, in: Harrison, P., Huchzermeyer, M., Mayekiso, M. (Eds.), *Confronting Fragmentation: Housing and Urban Development in a Democratic Society*. University of Cape Town Press. Cape Town, pp. 88–105.

Oldfield, S., Greyling, S., 2015. Waiting for the state: a politics of housing in South Africa. *Environment and Planning A*. 47, 1100–1112.

Paasche, T.F., Yarwood, R., Sidaway, J.D., 2014. Territorial tactics: the socio-spatial significance of private policing strategies in Cape Town. *Urban Studies*. 51, 1559–1575.

Pillay, A., Naudé, W.A., 2006. Financing low-income housing in South Africa: borrower experiences and perceptions of banks. *Habitat International*. 30, 872–885.

Reed-Danahay, D., 2005. *Locating Bourdieu*. Indiana University Press. Bloomington, Indiana.

Saborio, S., 2013. The pacification of the favelas: mega events, global competitiveness, and the neutralization of marginality. *Socialist Studies*. 9, 130–145.

Schinkel, W., van den Berg, M., 2011. City of exception: the Dutch revanchist city and the urban homo sacer. *Antipode*. 43, 1911–1938.

SERI, 2016. *Edged Out: Spatial Mismatch and Spatial Justice in South Africa's Main Urban Areas*. Socio-Economic Rights Institute of South Africa. Johannesburg.

Smets, P., 1999. Housing finance trapped in a dilemma of perceptions: affordability criteria for the urban poor in India questioned. *Housing Studies*. 14, 821–838.

Smets, P., 2004. *Housing Finance and the Urban Poor*. Rawat Publications. Jaipur.

Smith, N., 1987. Gentrification and the rent gap. *Annals of the Association of American Geographers*. 77, 462–465.

Smith, N., 1996. *The New Urban Frontier: Gentrification and the Revanchist City*. Routledge. London.

Smith, N., 2002. New globalism, new urbanism: gentrification as global urban strategy. *Antipode*. 34, 427–450.

South African Revenue Service Legal and Policy Division, 2006. *Guide to the Urban Development Zone Tax Incentive*. South African Revenue Service. Pretoria.

Statistics South Africa, 2012. *Census 2011 Statistical Release (No. PO301.4)*. Statistics South Africa. Pretoria, South Africa.

Swanson, K., 2007. Revanchist urbanism heads south: the regulation of indigenous beggars and street vendors in Ecuador. *Antipode*. 39, 708–728.

Teppo, A., Millstein, M., 2015. The place of gentrification in Cape Town, in: Lees, L., Shin, H.B., Lopes-Morales, E. (Eds.), *Global Gentrifications: Uneven Development and Displacement*. Policy Press. Bristol, pp. 419–440.

Tissington, K., 2011. *A Resource Guide to Housing in South Africa 1994–2010: Legislation, Policy, Programmes and Practice*. Socio-Economic Rights Institute of South Africa. Johannesburg.

Tomlinson, M., 2006. From 'quantity' to 'quality': restructuring South Africa's housing policy ten years after. *International Development and Planning Revue*. 28, 85–104.

Tomlinson, R., Beauregard, R.A., Bremner, L., Mangcu, X., 2003. The postapatheid struggle for an integrated Johannesburg, in: Tomlinson, R., Beauregard, R.A., Bremner, L., Mangcu, X. (Eds.), *Emerging Johannesburg: Perspectives on the Postapartheid City*. Routledge. New York, pp. 3–20.

Turok, I., Robson, B., 2007. Linking neighbourhood regeneration to city-region growth: why and how? *Journal of Urban Regeneration and Renewal*. 1, 44–54.

van Holt, K., Langa, M., Molapo, S., Mogapi, N., Ngubeni, K., Dlamini, J., Kirsten, A., 2011. *The Smoke that Calls: Insurgent Citizenship, Collective Violence and the Struggle for a Place in the New South Africa. Eight Case Studies of Community Protest and Xenophobic Violence*. Centre for the Study of Violence and Reconciliation and Society, Work and Development Institute, University of the Witwatersrand. Johannesburg.

Viruly, F., Bertoldi, A., Booth, K., Gardner, D., Hague, K., 2010. *Analysis of the Impact of the JDA's Area-Based Regeneration Projects on Private Sector Investments: An Overview*. Johannesburg Development Agency. Johannesburg.

Winkler, T., 2009. Prolonging the global age of gentrification: Johannesburg's regeneration policies. *Planning Theory*. 8, 362–381.

Zack, T., Bertoldi, A., Charlton, S., Kihato, M., Silverman, M., 2009. *Draft Strategy for Addressing Blighted Medium and High Density Residential 'Bad Buildings' in Johannesburg: Working Document for Discussion*. City of Johannesburg. Johannesburg.

3   The contradictory praxis of regeneration

Introduction

In this chapter I present a sociology of inner-city housing developers. Building on my analysis of the motivating logics and agendas which are framing the regeneration project, I focus on the accounts and perspectives of property developers and housing providers, and detail the ways in which they narrate their experiences, ambitions and actions in the inner-city. Their accounts show that they are deeply influenced by the social and spatial context in which they are operating. As a consequence, their habitus reflects a combination of developmental goals and practices as well as commercial concerns and market-based approaches to regeneration. The approach adopted here not only presents a new theoretical account about the relationship between space and social activity, but also offers a new perspective on urban regeneration. Just as Bourdieu attempted to produce a sociological explanation of economic action, this analysis offers a socio-spatial account of regeneration and the actors who drive it. It examines the effects which a regeneration agenda which is attempting to both provide housing for lower-income households and also generate commercial returns has on their practices. Thus, those who are shaping the inner-city and driving the regeneration process are not taken as representatives of capitalist imperatives or a market system which operates with its own inescapable logics, nor as part of a relentless march towards global gentrification. Rather, they are engaged with as complex social actors whose dispositions, views, interactions and ambitions are products of a conflicted, varied social order.

Analysis in the chapter pays careful attention to the spatial dynamics at play in the inner-city, and demonstrates how developers' actions are produced by their encounters with the particular spatialities of the inner-city. They are not only powerful players who are able to take and make place (although they are able to intervene in the physical and social realities of the area in significant ways), but are also contextually embedded actors whose dispositions and practices are shaped by the geographic setting in which they find themselves. Intervening in the inner-city means having to engage with its complex, messy logics and realities, and these realities, it will be shown, become absorbed into the ways in which housing providers understand and act in the space – i.e. their spatial praxis. The chapter further articulates the concepts 'spatial capital' and 'spatial habitus' and applies them to an empirical case. Drawing on ideas articulated by Lefebvre, I regard space as produced, dynamic and home to multiplicity, whilst I work with Bourdieu's understanding of habitus as a deeply ingrained, socially learnt disposition and capacity for relating to and acting in the world.

Working with the concepts 'habitus' and 'capital' presents some methodological challenges. Habitus is not readily verifiable or observable; it is a theoretical, hermeneutic device introduced by the researcher in order to identify, describe and analyse certain features of and actions within a social setting (Alvesson and Sköldberg, 2009; Burawoy, 2012). In conducting research, I did not set out to look for or identify a particular habitus, but rather developed ideas that a collective habitus exists amongst housing providers by being alert to recurring thoughts, expressions, affective states and actions which interviewees shared with me. Expressions or accounts which were repeated by numerous interviewees form the basis of what I identify as a shared habitus – i.e. 'a unity of style, which unites the practices and goods of a single agent or a class of agents' (Bourdieu, 1998, p. 8). As more interviewees used similar language or spoke about particular issues in recurring terms, it became apparent that a shared, collective way of looking at certain topics or subjects circulated amongst them. Given the predominance of critiques focusing on neoliberalism as the defining discourse in contemporary society, only the neoliberal/entrepreneurial habitus was a pre-identified/assumed category. The other aspects which constitute the habitus described in this chapter were identified inductively, as they emerged, sometimes unexpectedly, through the research process.

Habitus is also a term used to describe a deeply felt, affective inner-state and 'enables links between individuals' inner emotional worlds and external social and structural processes' (Reay, 2015, p. 22). Thus, during interviews I came to pay close attention to moments when interviewees became animated or emotional. In these cases, they were deemed to be expressing ideas or feelings that are deeply felt and intrinsic to their ways of relating to and working in the inner-city. They are, therefore, deemed to be constitutive of their motivating, guiding habitus. Lastly, habitus is also generative, a set of dispositions which translates into practice. Therefore, the effects of the habitus observed during interviews are made palpable through the ways spaces are made and claimed. Habitus in this sense is spatialized, for instance through decisions about where offices are located, the types of physical activities which take place in realising urban regeneration and the material interventions which are made into the built environment. I could personally perceive and experience all of these for myself, and thus speak to the veracity of the claims being made in interviews.

Entrepreneurial urbanism and the pioneering habitus in inner-city Johannesburg

Bourdieu advocated a form of analysis that regards social action as taking place within distinct (but overlapping) fields. Within each field, for example the art world (Bourdieu, 1984), education institutions (Bourdieu and Passeron, 1990), networks of consultants and policy-makers (Lingard et al., 2015) and even the professional boxing circuit (Wacquant, 2011), actors attempt to perform the identities and signs of distinction which are valued. To do so, they mobilise various forms of capital (material, symbolic, cultural) which allow them to achieve social dominance. It is possible to consider the city or neighbourhood as a field, as a setting in which diverse groups compete with one another for dominance in and over urban space (Isin, 2008). Some urban sociologists have already engaged in such analysis, as they focus on the practices and processes through which, for example, middle-class residents assert their claims to belong in gentrifying residential neighbourhoods, whilst simultaneously excluding or denying the claims of others (Benson and Jackson, 2013; Jackson and Benson, 2014). These selective strategies and forms of distinction allow middle-class residents to perform and establish identities for themselves as well as for their neighbourhoods, and thus aid them in the process of claiming space (Watt, 2009).

Within inner-city Johannesburg, property developers and housing providers can be regarded as a particular group who are working hard to assert their influence over the field. Middle-class residents in the aforementioned studies tend to share educational backgrounds, lifestyle preferences and professional status. Whilst property developers in Johannesburg are a relatively disparate group, including people of varying ages and with different educational and professional backgrounds, they can be considered to be part of a collective milieu for numerous reasons. Firstly, their positions as property owners give them shared economic interests and class identities, and also ensure that they have common ambitions for the area. Secondly, they have close professional relationships, with institutional representation taking place through the Johannesburg Property Owners and Managers Association (JPOMA). Although this influential lobby group does not solely represent inner-city housing providers, the current and former chairs are representatives of affordable housing companies based in the inner-city. Their institutional capital has been further boosted since the Democratic Alliance's electoral victory in the city. Given that the pro-business, libertarian mayor has made the inner-city a central focus, private developers working in the area have emerged as a highly influential constituency for him. Within the administration's first year in office, several high-profile meetings focusing on the inner-city were held between the City's executive team and private housing developers, and private developers have also been given opportunities, through both formal and informal channels, to influence policy and shape governance agendas.

Inner-city housing developers also share racial and gender backgrounds, as the majority of people running housing companies (with a few notable exceptions discussed later in this chapter) are white men. This does not necessarily mean that they share all life experiences and outlooks, but in a heavily racialised and gendered society like South Africa, it does point to a relatively strong basis for commonality and identification. Furthermore, three of the largest housing companies are currently run by the original founders' sons, and some employees of established housing companies have left to start their own, illustrating how close connections, institutional and cultural memories and practices are shared by actors within this field. They can, then, be regarded as sharing a particular form of habitus – a set of socially learnt, embodied and experiential dispositions, values and ways of being. The key concerns of this chapter are to outline what the motivating logics of this habitus are and how they contribute to shaping the forms regeneration is taking. As habitus is not only inculcated and socially learnt, but is also generative, the habitus being examined in this chapter matters because it is influential in producing space and affecting the lived realities of the inner-city.

Valorising urban entrepreneurialism

In a Bourdieuian conception, capital is multi-faceted, and refers to the different sets of socially valued attributes, characteristics and commodities which actors are able to mobilise to accumulate wealth and/or prestige and enhance their positions in the social hierarchy (Bourdieu, 1984, 1986). In a place-based perspective, spatial capital, then, is the ability to deploy various forms of capital and, in doing so, actively reshape physical spaces (Centner, 2008). As with the other forms of capital, exercising spatial capital requires an understanding and mastery of the prevailing tropes, mores and discourses which circulate in a specific society and geographic setting. In Johannesburg's inner-city, the form of capital and distinction which has become naturalised and which actors have to embody to achieve positions of dominance can be captured in the term 'the neoliberal-entrepreneurial habitus.' This refers to the set of dispositions and actions which embody the values and qualities espoused by neoliberal ideology, including belief in competition; trust in the market as the best mechanism for regulating society and solving social problems; a disavowal of reliance on the state or welfare; and an overtly economistic view of the world (Bourdieu, 2005; Comaroff and Comaroff, 2012). Because the regeneration project has been conceived broadly in terms which promote economic competitiveness, seek to reduce the role played by local government and place the impetus for regeneration in the hands of the private sector (Mosselson, 2017; Murray, 2008; Winkler, 2009), the field has been defined as one in which entrepreneurialism, profiteering and abilities to find privatised solutions to urban problems, particularly crime and violence, are valorised.

Housing providers gain significant amounts of spatial dominance and social and cultural capital through their abilities to enact the neoliberal, entrepreneurial habitus. As the previous chapter outlined, local government's regeneration strategies hinge on the private sector and their renewed interest in the inner-city. Governance activities are geared towards boosting private businesses' interest and investment in the area. Consequently, public interventions are measured against this goal. For instance, a report assessing the impact of the Johannesburg Development Agency's (JDA's) inner-city infrastructure projects commends the interventions that have been undertaken thus far. The authors of the report note with approval that, where the JDA has made infrastructure investments, vacancies have decreased, property transactions have increased, and property and rental values have improved. They thus come to endorse the approach to regeneration adopted by local government and deem the strategy to be a success (Viruly et al., 2010). In a clear sign of the symbolic power granted to private businesses and developers, the evaluation is not based on the extent to which poor people's access to housing is expanded or the ways in which these interventions affect local communities, but rather on the way in which investors and the market react. This means that private businesses become the arbiters of government interventions and actions, and the success of policy is measured according to their responses. The prevailing social order is thus one which clearly valorises entrepreneurial approaches to urbanism. Indicating how this disposition is shared by multiple actors in the field, the head of the City of Johannesburg's social housing company, JOSHCO, makes similar assessments. He declares that regeneration is succeeding since 'Now the conversation is about people wanting to buy; they don't want to get out, they want to get in to the inner-city property market.'

Taming the inner-city, exercising spatial capital

The prominence of business principles in government strategies and approaches to regeneration mean that being able to act entrepreneurially and take risks are key qualities which housing providers are required to possess. The bulk of private businesses and commercial investment firms continue to think about the inner-city in terms of abandonment, devastation and capital flight. However, those who are active in the housing sector do not share the broader perception, which translates into a significant amount of cultural capital for them in the field of the inner-city. Developers who entered the housing market in the early stages of regeneration are accorded a great deal of respect by their colleagues, competitors, city officials and employees from finance agencies. For instance, according to an employee of the state-owned National Housing Finance Corporation (NHFC), there was no meaningful activity in the housing sector in the inner-city before AFHCO, currently the largest housing provider, approached them with a proposal to purchase and refurbish buildings. He reflects on the early phase of the NHFC's relationship with AFHCO as follows:

> Prior to that the inner-city was almost going the other way, nobody wanted to [get involved], it was red-lined, banks refused to finance them and the buildings remained derelict. There was nothing significant that was taking place.

This sense that nothing meaningful was happening accords the early developers a large amount of authority and defines the type of urban regeneration currently being pursued as the only type worth mentioning. A younger employee of the Johanensburg Housing Company (JHC), who joined the company relatively recently, speaks with admiration for 'the guy who goes out first in that chaotic, crazy environment' and describes the first people to purchase properties and renovate them as low-income housing as 'the pioneers who set the way.'

The imagery of urban pioneers has been associated with gentrification and class change, particularly in American cities, and is generally used to romanticise and legitimise the violent displacement and class conflict which gentrification brings about (Smith, 1996). As the quotes above illustrate, a pioneering discourse is also present in Johannesburg's inner-city, but takes its meanings according to the particular nature of the social order and spatial setting. As Lingard et al. (2015, p. 25) emphasise, the forms of habitus which emerge in given fields are 'always vernacularized in particular ways [and] mediated by local histories, cultures and politics.' Thus, being pioneering in Johannesburg's context entails taking control over space through bravery, grit and an ability to take a violent, revanchist approach to dealing with potential threats and criminals. This is most effectively done by applying market-based approaches to urban management and security. The habitus which is valorised and translated into dominance in the inner-city thus combines pioneering, frontier spirit with an entrepreneurial perspective and the two together imbue developers with abilities to exercise spatial capital. As one for-profit developer aggressively declares,

> [developing housing in the inner-city is] not a game for sissies [a derogatory term for weak or cowardly people, particularly men, who are deemed to be inadequately masculine]; it's hands-on management. Dealing with the issues and the environment we operate in, it requires certain types of individuals.

Another also urges, 'If you don't have the strength and capacity for a very hands-on, involved approach, stay out!'

Therefore, in this field, spatial capital includes and is earned by the appetite to confront violence and utilise force, predominantly through private policing. Demonstrating this, one of the earliest inner-city developers explains the strategy his company adopted for dealing with building hijackers as follows:

> You've got to send in your security company with a task team, you know, 20, 30 guys and go and physically take them and throw them out because in those days if you relied on the police, they just wouldn't bother to pitch up 'cause quite a lot of the time they were part of the gang.

By confronting violence and mobilising their own, entrepreneurial forms of force, they thus exercise as well as gain spatial capital and earn the ability to take and make place. This ability has been crucial to dictating the types of spaces which are produced, as a once-chaotic, violent environment has now been made stable and is marked by the presence of private security personnel and surveillance equipment. Like City Improvement Districts, residential spaces of the inner-city reflect the expansion of privatised urban management and security provision solutions, and become spaces through which social order can be enforced and economic value can be extracted. Whilst rendering the area more governable and liveable, they also produce new forms of inequality and exclusion, which the subsequent chapters will highlight.

Envisioning redevelopment

However, pioneering housing developers did not only possess a habitus which helped them claim urban spaces and bring them under control; they also approached the area with outlooks which allowed them to formulate ambitions and see potential for hope and progress in it. The CEO of a for-profit housing company explains that when it comes to regeneration in the inner-city

> that entrepreneurial spirit is critical, to be able to see value, to have vision, to be able to see, well, this shithole where everything's destroyed and it's been stripped can be turned back to a functioning building.

Early developers and those who continue to be involved in upgrading the area do not subscribe to the doxa assumptions circulating about the inner-city, which continue to paint it as a place of failure, despair and intractable problems. According to one, the initial group of investors in the inner-city 'didn't see the risks in the same way' as the majority of people in the business community. A more recent participant in housing development reflects on the incredulity with which people in his social circle responded to his decision to invest in the inner-city. He relates that 'everyone told me I was nuts [crazy]!' He attributes people's alarm and scepticism to the enduring racism which still prevails in many sections of white South African society and which continues to equate black urbanisation with decay and disorder. Frustrated, he exclaims, 'Most white people are scared of Hillbrow; it's just prejudice.'

Although most housing providers in the inner-city, including the interviewees quoted above, are white, they have broken away from the doxa assumptions about race which continue to circulate in South Africa. Their pioneering quality was thus not only an ability to see economic opportunity in the inner-city and mobilise private forms of violence, but to appreciate the racial transition which the area has undergone. Crucially, they were able to discard attachment to the past, when the inner-city flourished but was a racially exclusive and discriminatory space, and do not harbour ambitions of returning to this period. In some instances, local government has been slower to adapt to the new uses and users which currently define the inner-city. Renewal plans developed by the City of Johannesburg have frequently been criticised for their enduring fixation with World City status and desire to recreate features of European urbanity in the inner-city (Bremner, 2000; Gaule, 2005; Sihlongonyane, 2015). There have also been ambitious and ill-suited plans formulated by some developers. For instance, one speculator hatched a scheme to renovate the iconic Ponte Tower and convert it into luxury apartments.[1] This plan failed and whilst the building has now been upgraded, it provides a much more modest offering. It currently serves as affordable housing, catering to people in the gap market with rents starting from R2100 for a one-bedroom apartment to R3700 for a three bed-room apartment, although a more luxurious option does remain – the two-bedroom penthouse apartment which rents for R4500.[2] In its current state the building is nearly fully occupied and is serving a purpose far better suited to the environment in which it is located.

'We're here, we live it every day': engaging with and embracing change in the inner-city

Responding to a changing context

In contrast to overly ambitious and out-of-touch speculators, consultants and policy makers, the developers who have been economically successful are those who have been able to recognise and respond to the inner-city's current form and function and who possess outlooks which embrace the racial and class transition which the area has undergone. Rather than regarding the area with apprehension, regret or gentrifying ambitions, they accept it for what it has become and the role it plays in the current South African context. One developer excitedly declares,

> We will never see the CBD of old, but there will be a new mix of service and retail sectors, not the heavy per square meter industries like finance, call centres and the like. A big portion will be residential but commercial elements of a different nature will also be present.

In speaking of the transition the inner-city has gone through, another private housing provider declares, 'I think it's all very exciting because of the strategic importance of Johannesburg, its location, the density of the population, the growing urbanisation and what the potential there is for work and earning a living.' This thinking is reflective of the more progressive strategies developed by local government and finance agencies seeking to capitalise on the possibilities the inner-city has to provide decent, centrally located housing. Housing providers' habitus, then, is aligned with and an outcome of the broader developmental and transformative agendas which circulate in post-apartheid Johannesburg. TUHF's CEO provides a good indication of the way those active in housing provision understand and come to embrace the changes the inner-city has undergone and appreciate its new dynamism, energy and significance in the post-apartheid landscape. Discarding the viewpoint which sees urban processes converging according to Anglo-European accounts and experiences, he explains,

> You know the inner-city decline in Joburg is not classic decline, like Detroit is classic decline, which is abandonment – Joburg's been the exact opposite … if anything, more people live and work in Joburg than ever before, and it's much more of a 24/7 type of city, and it's much more Afrocentric rather than Eurocentric, and the people and businesses who have left Joburg aren't

going to come back and the people who stayed probably aren't going to leave.

Because of these processes of change, the inner-city is a spatial reality which produces novel, sometimes improvised and clandestine, and always dynamic and highly charged forms of urbanity and survival (Simone, 2004). It also thus produces urban experiences, identities and habitus which are constituted in and by this diversity. For instance, one developer who lived through Johannesburg's segregated 'European' iteration and vividly recalls riding the tram into town as a child to visit the doctor, now points out that Johannesburg 'will always be an African city with culture, vibe, traditions.' For it to be African, however, does not mean that it is proliferated by dysfunction, informality and disorder; he points out that the people who are in the inner-city today also 'come to see gynies [gynaecologists], optometrists, dentists' and want 'a clean, safe place to raise families.' At the same time, people also come into the inner-city to buy wholesale goods which they will take back to their countries of origin to trade; the inner-city is thus, according to this interviewee, a 'multi-national, multi-social environment' and 'the shopping Mecca of Africa.'

Thus, housing providers have been able to embrace and draw encouragement from the changes which the area has undergone. This can be seen as a 'worlding' of their habitus (Ong, 2011, p. 13), as South African cities are re-situated in a wider world beyond European modernity. Many people in other sectors of South African society have failed to adjust their outlooks and remain intensely xenophobic and hostile to African migration (see Hassim et al., 2008; Misago et al., 2010; Neocosmos, 2010). In contrast, housing providers and property developers in the inner-city are far more adaptable and pragmatic. As one for-profit housing provider and property manager notes,

> We recognise there's a very large population of foreign Africans in Joburg and we find it useful to work with them because there's obviously significant demand and a large proportion of them. And also, we find them reasonable payers because obviously a lot of them are here to work, they're here because they're earning an income and they need accommodation.

Another developer explains that 'in the CBD-proper, in the east, you see a lot of Ethiopians, foreigners, the signs aren't even in English. We cater for those people.' He also highlights the buildings on the eastern end of Jeppe and Pritchard Street in which Somali and Ethiopian traders have established businesses and states emphatically that 'the commercial element is fantastic!' Creative leasing and retail strategies have also become commonplace. The interviewee quoted above points out

> there's also interesting trends with retail and commercial practices [in the CBD], with a lot of African trading where you get Ethiopian or Ugandan traders that take head leases on retail space and they sub-let garages or small

spaces to a variety of traders, which is quite interesting and quite common now in the inner-city.

Similar strategies have been met with hostility in other contexts, including supposedly diverse, cosmopolitan cities such as London (see Hall, 2015), but are generally tolerated or even supported in Johannesburg. By being less resistant to adaptations and creative uses of urban commercial space, property owners demonstrate their abilities to adjust to the area's changing nature and newly constituted diversity and multiplicity.

This embrace of the changes in the inner-city and its new type of population can be seen as a new form of habitus – a different schema of perception and appreciation which is able to move away from a fixation with the city's past or with reinventing it in line with world-class aspirations. Actors who are influenced by this habitus are far more readily pragmatic, accepting and able to see potential in the new urban situation. Their habitus is thus a reflection of the post-apartheid commitment to social redress and concern for the wellbeing of the inner-city and its population. It also reflects the diversity, worldliness and cosmopolitan nature of the area, and can be seen as a product of its spatial and social realities. Housing developers are not only working to create spaces of capitalism and further valorise the built environment; they are also responding to the needs and complex social realities of the area, which have come to shape their practices.

Place-based capital

The process of accepting change is a process of acquiring and coming to exercise spatial capital. In Bourdieuian interpretations, possessing capital means being able to successfully navigate particular social milieus or fields, and being able to embody the forms of distinction which are valorised within them. When viewed from a perspective which takes lived space and geographies seriously, capital is also determined by knowledge and awareness of the particular realities and social currents which are present in and characteristic of particular spaces. Spatial capital is thus knowledge and understanding of how a space functions, an ability to successfully navigate and be at home in a particular space, as well as the ability to act in and have influence over it. Significantly, the latter ability or power is contingent on successfully developing and inhabiting the former two aspects. Housing providers' experiences bear out this claim. There is a growing body of work which tries to understand the motivations and practices of property developers, who are a category of actors who possess significant power and abilities to alter urban landscapes (see for example Coiacetto, 2000; Kriese and Scholz, 2012; Ley, 2003; Zheng, 2013). Whilst this work has helped contribute to understandings of developers' networks, the ways they understand their roles, and the ways different national or regional contexts shape approaches to development, it largely lacks a spatial perspective, and does not account for the contingency, fluidity and adaptability of developers' practices. Most importantly, it does not theorise the ways in which experiences in space shape practices.

Developers are still largely presented as powerful actors who are able to exercise domination over space, in relatively straight-forward, linear ways.

However, as the case of regeneration in Johannesburg demonstrates, intervening in urban space is in no way a linear process through which interests, visions and agendas can be seamlessly imposed and realised. Rather, it is a constant process of adaptation and acquiring new spatial understandings and competencies. Successful housing providers are those who have been able to embed themselves in the realities of the inner-city and come to appreciate it in new and nuanced ways. As the above discussion illustrates, they have vision and an ability to recognise the transitions the area has undergone and to embrace the developmental potential the area has. They have a much closer engagement with its realties, and thus a greater understanding of its possibilities and positive aspects. All of the housing companies who have been commercially successful situate their offices in the inner-city, where their businesses and tenants are located. This ensures that they are familiar with the daily issues and encounters which arise in the area. As one developer explains,

> There's a reason why our offices are *here, in town*; there's a reason why AFHCO's offices are there, in town. That is because you don't want to run this business from an office in Rosebank [an upmarket commercial and residential suburb]. The moment you do that, you lose control of the business and you find that all kinds of things go to wrack and ruin.

Similarly, the employee at JHC who oversees the company's day-to-day operations and management tasks emphasises,

> If you're talking management in the inner-city you *cannot* have remote control management, you've got to have someone on-site there, living with the people, interacting on a daily basis, giving them service, hearing rumours and stuff and reacting to it before it becomes a problem.

This ability to engage with and understand the fabric of the inner-city is emphasised by one housing provider, a black woman, who used to be a domestic worker but now owns and manages a building providing student accommodation in the eastern part of the inner-city. She argues that her positionality, as someone who once had to live in, traverse and negotiate the inner-city, gives her an added advantage when it comes to doing business and providing housing. Comparing herself to two competitors, who are white businessmen who, according to her, 'grew up in property,' she explains that her life experiences have provided her with knowledge and competencies unavailable to them:

> because they were driving cars, I am walking. Now who is the better person? It's the one who's walking! Because you know: 'On this corner there are tsotsis [muggers], I mustn't go up there …' They are always driving past,

they don't know what's happening out there, so I'm the one who knows everything better!

She relates that even though they are competitors, they phone her and ask for advice about managing their property (demonstrating the close collaboration and shared project which housing providers in the inner-city have). Here it becomes apparent that traditional forms of cultural capital (being from a business background) and institutional capital (higher education qualifications) are not sufficient as they do not translate into spatial competency in and knowledge of the inner-city. Rather, it is the developers like her, who are able to come to grips with its messy, sometimes dangerous realities, who are in favourable positions and thus possess spatial capital. They are able to do this by situating themselves in its everyday contexts and having direct experiences of it, through walking its streets, locating offices there and dealing directly with the day-to-day management of the area. Through these located, place-based activities they come to possess intricate knowledge and understanding of the area and have what Wacquant (2011, p. 82) would describe as 'corporeal dispositions.' As one developer asserts, 'We're here, we live it every day.'

This point of view and experience is drastically different from many of Johannesburg's residents' relationships with the inner-city. For many, black and white alike, it continues to be a site of fear, a place to be avoided at all costs (Murray, 2011). Others have simply distanced themselves and disengaged from it. For instance, Duca describes a 'complete lack of interest in Johannesburg as a city' amongst residents (representing a variety of races) living in a golf estate in Roodepoort, an expanding commercial and increasingly residential suburb in the southern district of Johannesburg (Duca, 2013, p. 198). She observes that these residents, who can see the inner-city's skyline from their estate, have no 'spatial knowledge' or experience of using the city (ibid.). TUHF's CEO echoes this observation when discussing the commercial financial sector and their continuing reluctance to finance developments in the inner-city. He reflects on how he perceives a prevailing disposition which is distant from, and thus ignorant of, the realities of the area. Angrily, he declares,

> I think most of these banks are owned and managed by people who don't live or work in and don't understand the inner-city. They live in the northern suburbs and their life-view is very different. And I think that they're not only owned and managed by that [type of person], a lot of the workers are that kind of person. It is absolutely beyond me!

In contrast, housing providers are intimately acquainted with the inner-city and the people who live in it. They experience and engage with it as a lifestyle and form of everyday life. To live the inner-city means embracing it as a socio-cultural milieu in and of itself. Living, running businesses and trying to bring regeneration in the inner-city all constitute a way of being in the world, a world with its own experiences, ambitions, goals, outlooks, forms of distinction, stratification, survival and measurements of success. It is thus a space as Lefebvre understands the term, in all its multiplicity, unpredictability and sensuousness (Schmid, 2008) and is consequently one which produces distinct forms of habitus which arise directly out of the experiences, perceptions, identities and lived realities which it makes possible.

Regeneration as spatial praxis

Housing providers who embed themselves in this world and live it every day thus come to engage with its complex and messy realities and actively work to find solutions and make improvements. Doing so imbues them with spatial competence or capital, as they become able to master the environment and adapt their practices to most effectively deal with it. The different types of capital which they possess translate into and inform their practices, and thus a spatial praxis develops. As a representative of the City of Johannesburg's housing department states, social and affordable housing providers have 'an innate knowledge of how the city functions. It is hands-on knowledge.' At the same time, this knowledge and experience effects their dispositions and outlooks, thereby producing a habitus which reflects and adapts to the contingencies of the space. Mobilising spatial capital, then, is not the ability to transplant any vision or aspiration onto a space, but the ability to formulate actions, ambitions and worldviews which are appropriate reflections of spatial dynamics. In inner-city Johannesburg, doing so requires abilities to engage with the area's everyday realities, including its informality, volatile social relations, multitude of users and uses and even its violence. Acquiring these abilities allows individuals to under-stand the ways the area has changed and to recognise that there are different people inhabiting it who have different needs and socio-economic circumstances. From this recognition, they are then able to formulate appropriate development plans. As an employee of JHC reflects,

> There was major panic and hysteria, but when the dust settles it's not all doom and gloom. The more we hold on to the past, the more we don't succeed. It's not the old CBD of banks, it's something different. The people who were able to see that are the ones who benefited and made a difference.

Housing developers' reactions to the upmarket Maboneng precinct provide good examples of the ways they have 'let go' and adapted their outlooks. Located in Jeppestown, a former industrial suburb on the eastern edge of the Central Business District (CBD), the Maboneng precinct represents the clearest instance of gentrification in the city. Middle-class residents and tourists flock to the weekly food market and visit art exhibitions and shops selling luxurious items. It attracts a cosmopolitan, multi-racial crowd and also offers one of the few safe spaces for LGBTQI people, making it a welcome addition to the social geography of the city. At the same time, it is also a largely contrived space which attempts to recreate the feel of trendy urban areas around the world (Nevin,

2014). It has a graffiti wall (which can only be used with the permission of the area's management), a skateboarding ramp (which is used by children from the surrounding low-income buildings, rather than urban skaters or 'guerrilla urbanists' (Hou, 2010)), and high-end restaurants and bars which local working-class residents cannot afford to frequent. It is celebrated by residents who live there for the creativity, spontaneity and unexpected encounters which it fosters, but is also valued for the security it provides and the way it detaches itself from the unruly, run-down, poor part of the inner-city it is located in (Walsh, 2013).

The precinct has received a great deal of media attention and celebratory features about it have even appeared in the international press. However, whilst it has captured public imaginations and responds to a form of habitus which is more closely aligned with international, upper-class consumer culture, the development is largely dismissed by housing developers, who are focused on providing for the existing low-income community. One for-profit developer compares his company to Maboneng, which he describes as 'not viable.' Instead of focusing on 'sexy stuff,' they rather provide 'safe, solid, basic accommodation' which is much more in line with the needs of the majority of people settled in the inner-city. Another developer echoes this position and reflects that 'there is the cool, arty regeneration, but that's an artificial slice of Joburg, and then there are people who have made it their home,' people who come from across South Africa and the rest of the continent in search of stability and better economic opportunities and who 'want the easiest, cheapest accommodation.'

When they reject Maboneng, developers are reacting to – and capitalising on – the situation that actually presents itself. Rather than mimicking forms of urbanism or gentrification which are popularised around the world, their reading of and engagement with the inner-city produces alternative ambitions and desires. In one respect this response is driven by financial concerns and, like the finance institutions discussed previously, adapts to the demands of the market. Unlike Maboneng, which is attempting to create a new urban lifestyle and thus a new market and form of demand, they capitalise on the demand which is already there. Housing providers' success in the inner-city is determined by them 'understanding the market' and making sure that their products are 'affordable and value for money.' They therefore gain spatial capital by recognising what the people inhabiting the area need – 'the easiest, cheapest accommodation' – and formulating appropriate plans to deliver it. Doing so generates success and economic capital, as there is a much larger market to serve, but to do so one must first acquire spatial capital and adjust one's worldview and practices accordingly. As one developer explains,

I think that if we see the likes of a Main Street Life [one of the buildings in the Maboneng precinct] now starting to spread out and start to take over what was previously affordable housing, I think we have reason to be concerned. But I don't think that will ever happen in totality . . . You know, there is a market for the young yuppie environment but it's not a huge market, whereas the need for affordable housing is massive.

In dismissing Maboneng, these developers are also claiming symbolic power and capital in the field. Competitions for distinction take place between classes, but also within them (Bennett et al., 2009; Benson and Jackson, 2013). Therefore, by highlighting the viability and suitability of their projects, in comparison to others which are more superficially appealing and exciting but less developmentally attuned, these developers are performing and asserting their spatial capital. Doing so is important, as it becomes an additional source of economic and cultural capital. Whilst Maboneng is heralded in the media, the efforts of lower-income housing providers generally go unnoticed in broader society. However, those who align their projects with developmental efforts are able to receive recognition from local government and, more importantly, financial support from local and international institutions. For example, the French government's development agency, the Agence Française de Développement (AFD), has extended significant finance to one of the largest affordable housing companies in recent years. Developers who are able to claim symbolic distinction therefore gain in prestige, institutional connections and affiliations, and economic resources.

Therefore, housing developers' outlooks and aspirations are not simply the products of their entrepreneurial spirits and visionary, pioneering qualities. They are, rather, reflexive reactions to the realities and social dynamics they encounter and responses to the demands of the field they operate in. They are also expressions of deeply held, socially inculcated beliefs or subject positions. Habitus, Reay (2004, p. 435) reminds us, 'can be viewed as a complex internalized core from which everyday experiences emanate.' It also consists of 'passions and drives' which motivate and are reflected in actions and practices (Reay, 2015, p. 12). In the context of urban redevelopment, this internalized core relates to and emerges out of the place-making activities which developers engage in, activities which, as Benson and Jackson (2013, p. 794) argue 'may also be generative of subjectivities.' One interviewee clearly demonstrates the deep-seated, interiorised and affective disposition which he has acquired working in the inner-city. When asked about how he understands or approaches his work, he becomes very emotional and passionate. Rather than offering a descriptive term, he offers up an embodied, active understanding of what urban regeneration is:

> Urban regeneration for me, it has to be in your fibre and your way of looking at things and if you don't have that positive outlook – *you have to have that* in this inner-city. If you're not interested in urban regeneration you're going to be very frustrated in this place because it's a constant focus.

Significantly, this lived practice is formed both through experiences in space, as well as through the spatial realities and possibilities inherent in the regeneration project. Because it is a project which takes place in a particular, defined physical space, and also has the alteration of physical spaces and materials as its core concerns, it is one which revolves around adapting perspectives and allowing oneself to be shaped by the space.

In the process of doing so, one also gains the ability to enact ideals and values in space, and translate habitus into praxis. Just as claiming a neighbourhood or residential identity is not just a state of mind, but a process which is 'actualised in place and on the person' (Benson and Jackson, 2013, p. 798), realising regeneration as a spatial praxis entails making material interventions into the built environment. Thus, when asked why the company chose to focus on the inner-city as a site for housing provision, the CEO of JHC not only refers to the developmental possibilities of providing centrally located housing, but also stresses the materiality of the work they are engaged in, and the ways in which the lived physical realities of the inner-city contributed to defining their aspirations for the space. She explains:

> If you look at the amount of people that live here and work here, *in absolutely horrendous conditions*, why not the inner-city?! So the focus was to try and do two things … it was to try and create quality units where people need it, within the market that we defined [i.e. households earning between R3500 and R7500 per month] and also play a role in urban regeneration. So that was very crucial to us and that was what drove our decision.

Another interviewee running a social housing company echoes this, and takes great contentment from the ability to be able to alter urban environments, and, in so doing, create better social circumstances for people. When asked why he chose to become involved in housing provision in the inner-city he relates,

> I decided I wanted to do something to build a better South Africa. It is very satisfying because with housing you can *see* the change, how buildings and environments change. It's not like with HIV and sexual behaviour change, where you can't see anything. But when you see an area change for the better but can still accommodate people, it is very satisfying.

Again, these narratives can be understood as internalised frameworks for apprehending and acting in the world, and are reflections of a form of habitus which emerges in space, but also comes to effectively make space too.

Torn between competing demands

Commercial imperatives

However, whilst housing providers operate in a field which valorises and fosters a habitus oriented towards the improvement of the inner-city and the developmental goals regeneration is able to fulfil, they are also active in and shaped by a field influenced powerfully by commercial practices. In addition to a pioneering habitus, business proficiency is highly valued and the most successful housing companies are those that have 'access to capital, entrepreneurial and business experience and expertise,' as one interviewee described. For instance, two of the largest companies are headed by people who the same interviewee described as 'big businessmen,' one of whom used to be the financial director of a large insurance company and another who previously worked as the managing director of a publicly listed corporation. The capital, skills and habitus which contribute to success in the business world are important and come to define the field of urban regeneration too. This does not only apply to for-profit companies, as social housing institutions also have to master these skills. Hence the head of JOSHCO proudly insists, 'We formulate our business model on normal real estate principles. We are a business, not a municipality.' Another smaller social housing company, which is part of a faith-based organisation, still operates as 'a lean and mean machine' and receives plaudits from its peers and government for being 'run on a very efficient business basis.'

Habitus is a socially learnt competence or mastery of the forms of presentation, expression and social action which are valued in a social field. Thus, Bourdieu's (1990, p. 66) explanation that habitus is a 'feel for the game' has been widely adopted and often repeated. In the case of housing providers, their actions are well-attuned not only to the new spatial reality of the inner-city, but also to the prevailing economic climate in which they operate. Thus, as the property market normalises and competition for properties and tenants becomes greater, developers are required to be adaptive and innovative. As a for-profit developer points out, 'Guys who are more efficient, competitive, will manage and others will fail.' In this way, the logic of the market becomes absorbed into their actions and ways of being, and abilities to negotiate unforgiving commercial terrain are essential sources of capital. The CEO of JHC, although she runs a social enterprise, asserts,

> The thing is, for a non-profit a lot of people have this idea that if you're a non-profit you can't do financial engineering, you shouldn't look at funding like businessmen – *you should*! Because that's *the only way*. For JHC, our development objectives are crucial, but we cannot achieve our development objectives if we do not make enough money to do so.

Thus, whilst transformative goals are central to the regeneration project, these have to be pursued within the constraints imposed by the market. Developers have to cater to a low-income population but also ensure that they are able to recoup their costs and charge rents which allow them to cover their maintenance and operating expenses. An employee of JHC notes that 'we don't cater to wealthy people with free resources.' Another private developer echoes this when he explains that 'there are real affordability challenges for that demographic [that rents in the inner-city], if you just think about the poverty challenge in this country, the housing challenge etc.' Thus, housing developers have to be cognisant of and adapt to these needs. They have introduced innovative practices such as offering cash incentives to existing tenants who refer new clients to them, providing one month's free rental to new tenants and, in the case of social housing, offering financial support and covering funeral costs if the main lease-holder dies. They also condone sub-letting, which enables tenants to divide the costs of rent between multiple households.

At the same time, expanding businesses, repaying loans and generating profits are essential. On top of this, the costs of electricity, water and property rates continue to rise, making providing cheaper rentals increasingly difficult. One developer explains, 'We're facing a battle now where rates are going up by 600% on the value of our property. The impact on our business is significant, an extra R400 000 a month.' Another points out that 'over the past four years, we've had a 164% increase in the cost of electricity [and] sewerage has gone up 80 or 90%.' To compound matters, because some buildings have been converted from former commercial or office space they continue to be charged commercial rates, which are more expensive than those charged for residential properties. Because social housing providers cater to people with lower incomes, they are even more pressurised by rates increases. As a senior employee at JHC explains,

> The biggest problem is our operating costs ... at the end of the day we don't get any discount because we're nice guys, we pay the same for electricity as any commercial landlord, and for rates, and for security guards, all of that stuff ... things costs the same whether your mission says you want to be below the market, it doesn't give you any discounts.

Thus the enforcement of user-pays principles and cost-recovery practices by national agencies such as ESKOM, the national electricity provider, and local government's strategy of using inner-city property taxes to raise its revenues are placing greater strain on the affordability of housing in the inner-city. The overburdened nature of the regeneration process becomes clear, as do the negative effects of a market-based approach to housing provision. Developers have to negotiate between the conditions of finance agencies, social commit-ments and the commercial imperatives which are imposed on them. They there-fore come to inhabit a habitus which reflects the contradictory social order they are acting within and are split between diverse goals for and approaches to regeneration.

These come to the fore when poor and vulnerable communities' claims to space in the inner-city are discussed. Housing providers express concern for them and regret evictions and displacement, but simultaneously defend their own claims to property and the role they are playing in improving the area. A for-profit developer indicates this clearly when he reflects on evictions he has been involved in. He veers between concern and anger, declaring, 'It truly breaks my heart, but when someone steals your car you don't say "Oh, shame," [an expression used in South Africa to convey sympathy] you get angry! And it's the same with property rights.' Similarly, another developer describes her dilemma in having to evict occupants from the building she purchased. Again indicating the powerful emotions which shape and are enacted in habitus, she recounts,

> It's actually hurting inside. Firstly that time, it was raining from Monday to Monday and when I'm at home they [the occupants facing eviction] were just in my mind, but I didn't know what to do. Because I tried to speak to them, explain, trying to reason with them. But they didn't want to understand, because they were not paying the rent, they were staying for free. So it's not easy to take someone who's staying for free and make her pay rent. So it was even hard for me to pass here because they were living on the street.

The habitus of housing providers and other actors involved in urban regeneration is pulled in two directions and reflects the fact that there are competing social orders and frameworks for organising and shaping social life in the inner-city. They are torn between the need to valorise space and convert it into more economically productive uses, as well as the need to use it to enhance social and spatial integration and the inclusion of low-income communities in the urban fabric.

Thus, as much as the desires to do good and make positive changes to the urban and social environment are deeply held convictions, they are also justifications and rationalisations which help developers reconcile their actions with the displacement and harm which redevelopment causes. They are, therefore, attempts to suture their cleft habitus. This is exemplified by a social housing provider, who captures the contradictions and conflicts which define inner-city Johannesburg and the habitus of those involved in regeneration efforts, when he muses

> I love [my job]. The stuff I used to be involved in was predictable. Now it is much more interesting. Who else will buy a building that's up to shit? ... Now I get to see people live in safe, clean nice accommodation. You forget about it because it's just a job, but some buildings – if I was more emotional I would burst out crying. It's so awful to see [the conditions some people live in] but I know when I buy the building it becomes safer, cleaner for people to live in. Sure, the existing occupants will have to move out ...

Racial divisions and distinctions

It is therefore clear that, although a social field is constituted by multiple actors engaging in and sharing similar experiences, worldviews, economic conditions and habitus, there are competing priorities and dispositions within a single field which push and pull actors in different directions. Fields are also characterised by distinctions, hierarchies and struggles for recognition and prestige *within* classes too (Bourdieu, 1984). Housing providers in the inner-city, as should be apparent by now, have a shared set of experiences and habitus, and thus can be thought of in collective terms. At the same time, however, within this milieu there are conflicts and competing forms of distinction and capital at play. As in all South African social interactions, race serves as a key point of conflict and marker of difference. As mentioned earlier, the majority of ownership positions in housing companies are occupied by white people, and the majority of them are male. The business backgrounds they brought with them to the affordable housing sector also meant that they brought established stocks of economic capital, and, in a world marked by enduring racial prejudices, social capital too. There is a small cohort of black developers who, particularly through the focused financial and business planning assistance provided by TUHF and the GPF, have been able to enter the field. However, they continue to complain about the prejudice they experience. Two developers recount the ways employees of commercial finance agencies regard black people with suspicion when they apply for finance, and continue to regard property development as a job for white people. For example, the developer who was previously a domestic worker shares her frustrations and recounts how, when she first seized on the idea of purchasing and redeveloping a building, none of the commercial banks were willing to take her proposal seriously. With anger, she recounts, 'Nobody believed in a black woman! They just saw us like "They must do the washing, cook, bear children."' Thus, despite her stocks of spatial capital, she bemoans the fact that financial institutions adhere to racial stereotypes and are only concerned with and recognise people's formal qualifications – their institutional capital – and are unable to acknowledge the different skills and resources people possess – their cultural and spatial capital which reflect and are valuable within the post-apartheid social order and changing urban context.

It therefore becomes apparent that, as much as new dispositions and spatial praxis are required and rewarded in the inner-city, there are also enduring racial and cultural distinctions and forms of stratification which are maintaining the prevailing racist structure of South African society. Although TUHF and the GPF make concerted efforts to support black entrepreneurs, the social and cultural capital which white developers possess allow them to maintain positions of dominance. Another black developer notes that black people, in addition to lacking the economic resources to enter the property market, generally do not have the connections (social capital) and footholds in the world of finance to source the deals which will allow them to invest. He complains that they are reliant on TUHF, whilst other white-owned companies with established track records are able to utilise their reputations and international networks to leverage more finance. This is borne out by AFHCO's success in securing finance from AFD. They were awarded this grant on the basis of their dominance in the inner-city market, which results partly from the forms of racial distinction and capital which they already possessed, and which ensure that patterns of skewed racial ownership and economic mono-polies (what some might call 'white monopoly capital') reproduce themselves.

There are some state-led efforts to address these imbalances. Despite the success of the Better Buildings Programme (BBP), it was discontinued in 2007, and replaced by the Inner City Property Scheme (ICPS). The BBP was criticised for using state resources to support white-owned companies, and it was decided that state assistance should be focused on supporting emerging black entrepre-neurs instead. Whilst the need for transformation of the economy and patterns of ownership are indeed dire, the difficulty has been that, as mentioned above, there are few black-owned companies who have the requisite capital and experience to enter the inner-city property market, even with state assistance. Thus, the ICPS has been much slower to deliver results, yielding only three projects since its inception. The previous chapter pointed to the overburdened nature of the regeneration process, which is again borne out in this case. Through the ICPS, property and housing in the inner-city are being envisioned as vehicles for supporting the growth of a black business class, as well as, and in some cases over and above, housing provision. The reliance on a market-driven paradigm, as well as the particular racial dynamics and transformation imperatives which characterise post-apartheid society, as well as the competing needs which regeneration is attempting to fulfil, again come to the fore, and are shown to play central roles in shaping, and at times disrupting, the character and outcomes of the process.

Conclusion: a contradictory, localised process

It therefore becomes clear that we cannot think of inner-city regeneration without keeping an eye firmly fixed on the competing imperatives, agendas and socio-political dynamics which define the post-apartheid context. At the same time, the particular spatial realities of the inner-city come to the fore as determining conditions which shape even the most dominant actors' disposi-tions and practices. Housing providers' habitus is split between the demands of the market and the imperatives of post-apartheid development and trans-formation. These dualities have arisen out of the contested and opposing ways in which regeneration has been framed and the contradictory effects they have when they are translated into social actions and spatial practices. Housing providers are shown to possess forms of habitus which reflect these contra-dictions and reproduce them in the spaces of the inner-city through the ways they translate into spatial praxis. The regeneration process is thus not only framed but is actively pursued in dualistic, contrasting ways.

At the same time, housing providers are not only influencing the inner-city and imposing a (contradictory) project onto it. The inner-city itself presents and produces a complex lived reality which actively shapes housing providers' outlooks and practices. In order to be successful and dominant in the space, they have to acknowledge and adapt to its realities. This requires being able to act entrepreneurially, embrace a pioneer spirit and employ the services of private security companies, but it also requires adopting developmental approaches to regeneration and a willingness to work with the African populations who now reside in and define the area. Through their close engagement with the area, housing providers' habitus has been adjusted and they have acquired spatial capital which allows them to formulate ambitions and approaches to regeneration which are appropriate for the diverse, dynamic space. This makes the practices and ambitions they adopt, and thus the regeneration process, vernacular. It is not one framed by ambitions arriving from or oriented elsewhere, and firmly rejects the imposition of a gentrification aesthetic. It therefore highlights the specificities of the context and the way in which it is absorbed into and shapes the practices of urban actors.

To frame the process as 'vernacular' is not necessarily to celebrate it. Rather, it is to point out the forms of contradiction, antagonism and conflict which are derived from the particular spatial and temporal moment, and which play out in particular ways. Cognisance of these competing agendas and practices is heightened by paying attention to the spatial practices which developers engage in, and by being alert to the extent to which space is not simply acted upon, but actually shapes dispositions and practices. The concepts 'spatial capital' and 'spatial habitus' aid us in doing so, and may be of utility in thinking through relations between space and subjectivity in other settings too.

Adding a spatial perspective to Bourdieuian theorisations about social action and reproduction helps avoid some of the pitfalls associated with this stream of thinking. By being alert to the dynamism, multiplicity and in-process nature of space – as both a material and social entity – research can remain alive to the possibilities and ways in which habitus is constantly shifting and developing. By doing so 'the seeming inevitability of reproduction' (Gale and Parker, 2015, p. 82) which Bourdieu's theories can often lean towards, can be discarded in favour of more varied, context-sensitive and nuanced appreciations of spatial praxis. Similarly, working with ideas of spatial habitus and spatial capital can help recover the sense of dynamism which Lefebvrian analysis should aim to achieve. Returning to the example used in this chapter, the ambitions which developers hold for the inner-city cannot be simply reduced to profit-seeking, extractive goals, nor are the types of spaces which they produce simply commodified, abstract spaces, although this is one of the results. Rather, we can see how spatial representations and practices are hybridisations of diverse socio-political currents and dispositions, which take shape in wider socio-political contexts, and then feed into the practices of developers and the types of representations and agendas they formulate. This too is a dynamic, lived process and needs to be regarded as one which entails a range of different forms of capital, experiences and habitus coming together. Through attention to these competing imperatives and lived realities, we can arrive at a more complete picture of how processes of spatial domination and construction take place, and more accurately account for the various factors which shape these.

Notes

1 This project is chronicled in Rehad Desai's (2010) documentary *The Battle for Johannesburg*.
2 http://mg.co.za/article/2012-04-20-pontes-fourth-coming-an-urban-icon-reborn.

References

Alvesson, M., Sköldberg, K., 2009. *Reflexive Methodology: New Vistas for Qualitative Research*. SAGE. California.

Bennett, T., Savage, M., Silva, E.B., Warde, A., Gayo-Cal, M., Wright, D., 2009. *Culture, Class, Distinction*. Routledge. New York.

Benson, M., Jackson, E., 2013. Place-making and place maintenance: performativity, place and belonging among the middle classes. *Sociology*. 47, 793–809.

Bourdieu, P., 1984. *Distinction: A Social Critique of the Judgement of Taste*. Harvard University Press. Cambridge, Massachusetts.

Bourdieu, P., 1986. The forms of captial, in: Richardson, J.G. (Ed.), *Handbook of Theory and Research for the Sociology of Education*. Greenwood. Westport, Connecticut, pp. 241–258.

Bourdieu, P., 1990. *The Logic of Practice*. Stanford University Press. Stanford, California.

Bourdieu, P., 1998. *Practical Reason: On the Theory of Action*. Stanford University Press. Stanford, California.

Bourdieu, P., 2005. *The Social Structures of the Economy*. Polity Press. Cambridge.

Bourdieu, P., Passeron, J.-C., 1990. *Reproduction in Education, Society and Culture*. SAGE. California.

Bremner, L., 2000. Reinventing the Johannesburg inner city. *Cities*. 17, 185–193.

Burawoy, M., 2012. Theory and practice: Marx meets Bourdieu, in: Burawoy, M., van Holt, K., *Conversations with Bourdieu*. Wits University Press. Johannesburg, pp. 31–45.

Centner, R., 2008. Places of privileged consumption practices: spatial capital, the dot-com habitus, and San Francisco's internet boom. *City and Community*. 7, 193–223.

Coiacetto, E.J., 2000. Places shape place shapers? Real estate developers' outlooks concerning community, planning and development differ between places. *Planning Practice and Research*. 15, 353–374.

Comaroff, J., Comaroff, J.L., 2012. *Theory from the South: Or, How Euro-America Is Evolving toward Africa*. Paradigm Publishers. Boulder, Colorado.

Desai, R., 2010. *The Battle for Johannesburg*. Uhuru Productions. Johannesburg.

Duca, F., 2013. New community in a new space: artificial, natural, created, contested. An idea from a golf estate in Johannesburg. *Social Dynamics: A Journal of African Studies*. 39(2), 191–209.

Gale, T., Parker, S., 2015. Calculating student aspiration: Bourdieu, spatiality and the politics of recognition. *Cambridge Journal of Education*. 45, 81–96.

Gaule, S., 2005. Alternating currents of power: from colonial to post-apartheid spatial patterns in Newtown, Johannesburg. *Urban Studies*. 42, 2335–2361.

Hall, S.M., 2015. Super-diverse street: a 'trans-ethnography' across migrant localities. *Ethnic and Racial Studies*. 38, 22–37.

Hassim, S., Kupe, T., Worby, E. (Eds.), 2008. *Go Home or Die Here: Violence, Xenophobia and the Reinvention of Difference in South Africa*. Wits University Press. Johannesburg, South Africa.

Hou, J., 2010. *Insurgent Public Space: Guerrilla Urbanism and the Remaking of Contemporary Cities.* Taylor & Francis. Abingdon.

Isin, E.F., 2008. The city as the site of the social, in: Isin, E.F. (Ed.), *Recasting the Social in Citizenship.* University of Toronto Press. Toronto, pp. 261–280.

Jackson, E., Benson, M., 2014. Neither 'deepest, darkest Peckham' nor 'run-of-the-mill' East Dulwich: the middle classes and their 'others' in an Inner-London neighbourhood. *International Journal of Urban and Regional Research.* 38, 1195–1210.

Kriese, U., Scholz, R.W., 2012. Lifestyle ideas of house builders and housing investors. *Housing Theory and Society.* 29, 288–320.

Ley, D., 2003. Artists, aestheticisation and the field of gentrification. *Urban Studies.* 40, 2527–2544.

Lingard, B., Sellar, S., Baroutsis, A., 2015. Researching the habitus of global policy actors in education. *Cambridge Journal of Education.* 45, 25–42.

Misago, J.P., Monson, T., Polzer, T., Landau, L.B., 2010. *May 2008 Violence against Foreign Nationals in South Africa: Understanding Causes and Evaluating Responses.* Consortium for Refugees and Migrants in South Africa and Forced Migration Studies Programme. Johannesburg.

Mosselson, A., 2017. 'Joburg has its own momentum': towards a vernacular theorisation of urban change. *Urban Studies.* 54(5), 1280–1296.

Murray, M.J., 2008. *Taming the Disorderly City: The Spatial Landscape of Johannesburg After Apartheid.* Cornell University Press. Ithaca, New York.

Murray, M.J., 2011. *City of Extremes: The Spatial Politics of Johannesburg.* Duke University Press. Durham, North Carolina.

Neocosmos, M., 2010. *From 'Foreign Natives' to 'Native Foreigners': Explaining Xenophobia in Post-Apartheid South Africa: Citizenship and Nationalism, Identity and Politics.* African Books Collective. Oxford.

Nevin, A., 2014. Instant mutuality: the development of Maboneng in inner-city Johannesburg. *Anthropology South Africa.* 37, 187–201.

Ong, A., 2011. Worlding cities, or the art of being global, in: Roy, A., Ong, A. (Eds.), *Worlding Cities: Asian Experiments and the Art of Being Global.* Blackwell Publishing. Chichester, Sussex, pp. 1–26.

Reay, D., 2004. 'It's all becoming habitus': beyond the habitual uses of habitus in educational research. *British Journal of Sociology of Education.* 24, 431–444.

Reay, D., 2015. Habitus and the psychosocial: Bourdieu with feelings. *Cambridge Journal of Education.* 45, 9–23.

Schmid, C., 2008. Henri Lefebvre's theory of the production of space: towards a three-dimensional dialectic, in: Goondewardena, K., Kipfer, S., Milgrom, R., Schmid, C., (Eds.), *Space, Difference, Everyday Life: Reading Henri Lefebvre.* Routledge. New York, pp. 27–45.

Sihlongonyane, M.F., 2015. The rhetorical devices for marketing and branding Johannesburg as a city: a critical review. *Environment and Planning: Economy and Space.* 47, 2134–2152.

Simone, A.M., 2004. People as infrastructure: intersecting fragments in Johannesburg. *Public Culture.* 16, 407–429.

Smith, N., 1996. *The New Urban Frontier: Gentrification and the Revanchist City.* Routledge. New York.

Viruly, F., Bertoldi, A., Booth, K., Gardner, D., Hague, K., 2010. *Analysis of the Impact of the JDA's Area-Based Regeneration Projects on Private Sector Investments: An Overview.* Johannesburg Development Agency. Johannesburg.

Wacquant, L., 2011. Habitus as topic and tool: reflections on becoming a prizefighter. *Qualitative Research in Psychology*. 8, 81–92.

Walsh, S., 2013. 'We won't move': the suburbs take back the centre in urban Johannesburg. *City*. 17, 400–408.

Watt, P., 2009. Living in an oasis: middle-class disaffiliation and selective belonging in an English suburb. *Environment and Planning: Economy and Space*. 41, 2874–2892.

Winkler, T., 2009. Prolonging the global age of gentrification: Johannesburg's regeneration policies. *Planning Theory*. 8, 362–381.

Zheng, J., 2013. Creating urban images through global flows: Hong Kong real estate developers in Shanghai's urban redevelopment. *City, Culture and Society*. 4, 65–76.

4 Urban management and security

Private policing, atmospheres of
control and everyday practices

Introduction

In this chapter I explore the effects the regeneration process is having on
public spaces in the inner-city, and examine the diverse meanings which
security, policing and safety come to have in this context. Following from
the previous chapters, I will show how the deployment of spatial capital
means being able to make and shape spaces, and that housing companies, in
partnership with the private security firms they have hired, do so in palpable
ways. Their dominance enables them to enforce regulation and forms of
social ordering and remake spaces in accordance with the entrepreneurial,
commercially focused goals they are pursuing. However, exercising spatial
capital means adapting to the particular conditions which determine a space,
and shaping practices in accordance with prevailing social–spatial logics.
Alongside coercive forms of policing, subtle processes of domestication,
community-building, place-making and affective, atmospheric regulation are
taking place which respond to the local populations' needs and desires and
attempt to make the inner-city a hospitable, inclusive environment. Again,
the process of making and shaping spaces is shown to avoid easy classification or description, and comes to take on vernacularised, multi-faceted
forms.

As the chapter unfolds, I will unpack the various logics which inform the
security and policing practices at work. It will be demonstrated that the process
encompasses a range of actions and forms of policing, each with their own
register of legitimacy and socio-cultural meaning. Crucially, the practices which I
outline, whilst in some cases bearing strong hallmarks of revanchist policing
(MacLeod, 2002; Smith, 1996) and having exclusionary outcomes, are spatially
situated, and therefore draw their logic and legitimacy from the realities which
prevail in the inner-city. They are therefore, in many cases, more subtle and
restrained than the revanchist practices which tend to be highlighted in academic
work focusing on the regulation and privatisation of public space, both in South
Africa and internationally. As such, they are often harder to notice, but also more
effective at bringing lasting change to the area; they are also harder to evaluate
and assess, and stand as ambiguous interventions. Whilst discussing broader

trends in the inner-city, the chapter predominantly draws on research conducted in the Ekhaya Neighbourhood in Hillbrow.

The 'precinct and project approach' and exercising spatial capital

Clustering investment

The early phase of the inner-city regeneration process was marked by what Beall et al. (2002) term a 'project-and-precinct approach.' Drawing on American models of inner-city upgrading and heavily influenced by consultants with direct experience in American cities, the City of Johannesburg inaugurated several high-profile regeneration initiatives in the early 2000s. These were geographically bounded and aimed at making interventions into the infrastructure of specific areas so that they could be better geared towards a range of predetermined economic functions. As such, the precinct approach is a form of placemaking which emphasises particular features of urban areas and aims to enhance the desirability, image and marketability of these spaces, ultimately with an eye to augmenting the value of land within them.

A precinct-based approach is an exercise in using spatial capital because it entails imposing particular visions and aspirations onto given spaces, thereby recreating them to serve dominant groups' desires and agendas. Precincts which have been established around the inner-city include the Newtown cultural precinct, the Diamond and Fashion Districts in the eastern segments of the CBD and the Ellis Park Sports Precinct in Doornfontein, on the eastern edge of the inner-city. These were all commissioned in the mid-2000s and were intended to stimulate and secure private investment in the inner-city (Viruly et al., 2010). These projects therefore respond to and produce particular imaginations of the areas they are imposed upon. They focus on visualising them as clean, economically productive, managed, marketable and lucrative business destinations. For instance, the Fashion District, when it was unveiled, was launched with promises to attract 'fashion-conscious suburban shoppers, downtown shoppers who work in the CBD, and tourists.'[1] Colourful public artworks, boutique shops and an outdoor amphitheatre and catwalk were all installed as key features used to distinguish the district from the surrounding area and give it its unique identity.

Similarly, the Ellis Park Sports Precinct, developed for the 2010 FIFA Football World Cup, aims to make Johannesburg a tourist destination which appeals to thrill-seeking, wealthy and adventurous travellers (Bénit-Gbaffou, 2009). The plans for the sports precinct have largely fallen flat, but in a telling indication of the types of users and inhabitants who the upgrades were aimed at, a local community sports centre was renovated as part of the upgrading process and, despite the fact that it was done under the auspices of preparations for the Football World Cup, resulted in a local soccer field being converted into a cricket facility (Bénit-Gbaffou et al., 2013). Soccer is the most popular sport in South Africa and holds particular appeal amongst the black majority; cricket, in contrast is a sport generally enjoyed by the wealthier white and Indian communities. In this example, it becomes clear that imaginations which sought to make the inner-city attractive to affluent and more 'desirable' classes of people were highly influential. Most importantly, these imaginations did not only operate in the conceptual realm, but materially reordered and produced the space, with unfortunate consequences for those who did not fit into the dominant visions.

Housing companies have seized on the precinct approach to renewal, as it helps augment their control over spaces, particularly the public areas surrounding their buildings. In the initial phases of upgrading, companies were successful in bringing individual buildings under control and making them safer spaces for their tenants, but the surrounding environment continued to be dangerous, threatening and difficult to manage. Thus, as City Improvement Districts (CIDs) gained popularity in commercial areas of the inner-city and elsewhere, housing companies too began to participate in them. In cases where CIDs weren't feasible, companies used the deflated property market to their advantage and bought up buildings in close proximity to one another. Doing so enabled them to cluster their efforts and resources, and was a way of augmenting their dominance over a given space. Map 4.1 shows the location of the Johannesburg Housing Company's (JHC) inner-city properties, and clearly demonstrates how they have been bought in clusters, thereby giving the company a marked presence in whole city blocks.

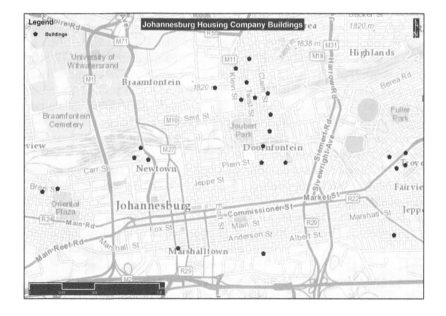

Map 4.1 Map of JHC building clusters.

Discussing this strategy, the company's Operations Manager explains how it enables property owners to concentrate resources and extend their reach into public spaces, thereby allowing them to shape areas in ways which they deem most desirable. As he explains,

> Our aim is to take that building and now make a difference in that area … The whole idea is for a building to affect the area that it's in, for us to drive that and how it affects it. It's easier when it's a group of buildings.

Other companies have also adopted this method. As an employee of one of the largest private companies points out,

> What you're doing when you have a cluster of buildings together, you *uplift* that *area* because of your investment in that area, compared to if you've got one here, one here, one there, it's more difficult to uplift that area. But if you've got three or four [buildings] around one another it's easier to uplift that area. Even if it's not your own you can clean it; if this is a building here and there's a building next door that's not yours and then [the next building is] yours, if the one in the middle is not cleaning his pavement, your guys will clean the pavement because he doesn't want his building to be bad; your two building managers will ensure that that area is clean and presentable.

Consequently, the inner-city has been divided up into different, informal precincts. Companies utilise distinctive branding, marking areas with their presence and creating discernible pockets of regeneration. The most notable example of this is the Ekhaya Neighbourhood, which has been established in the southern section of Hillbrow. The precinct came into being as an informal CID and is classified as the first, and at this stage only, Residential City Improvement District (RCID) in Johannesburg (Peyroux, 2012). Its genesis came when two companies, JHC and Trafalgar, who had buildings diagonally across from each other in the same street, agreed to cooperate on security matters. They decided that, in addition to having guards working inside their buildings, they would station them on the streets outside each building too. Because of their positioning, the guards were able to keep watch over the entire street, and quickly became a powerful deterrent to potential criminals. The success of this initial experiment motivated other landlords with properties in the area to come on board, and soon an informal association was formed. Before a recent change in legislation,[2] which has re-classified all CIDs as voluntary associations, formal CIDs required the consent of 51% of property owners in a designated area; there are still a number of derelict buildings and slumlords in the area which grew into the Ekhaya Neighbourhood. Thus, the companies involved were unable to get majority consent, and instead pooled their resources to create a voluntary association. Like other CIDs, this association pays for cleaning and maintenance services, has engaged in some forms of branding and placemaking and hired the services of a private security company.

The upside of the Ekhaya Neighbourhood and other areas where precinct strategies have been adopted, such as Legea Lerona in Berea, has been the focused upgrading and intensive policing and maintenance activities which have taken place. The downside, however, is that the process continues to be driven by property development, and, consequently, areas which haven't been attractive to or conducive for a precinct strategy remain in poor conditions. Furthermore, when companies have assumed control of precincts or swathes of inner-city territory, they have been able to exercise control in sometimes (but not exclusively) revanchist ways. The previous chapter pointed out that the habitus and forms of capital which prevail require investors to be able to formulate privatised approaches to securing inner-city spaces, and that they have most frequently accomplished this by hiring private policing services. By doing so, companies are able to police areas and secure their interests, and also realise their visions for space. This has had negative effects on those people and populations who do not fit into the visions and desired orders companies are using their spatial dominance to create.

Regulating space, producing populations

As private security has been introduced in the Ekhaya Neighbourhood, a 'zero tolerance' approach to policing and public space management has come into force. As in other CID areas in South Africa, and cities across the globe where private policing has extended into the public realm, populations of homeless people, young unemployed men and people engaged in informal economic activity have come to be the focus of policing efforts (see Cook, 2010; Didier et al., 2012; Eick, 2012; Lippert, 2012; Miraftab, 2007; Paasche et al., 2014). The company charged with policing the neighbourhood goes by the name Bad Boyz. Their guards are stationed at the entrances to buildings which have contracted their services, and they also conduct patrols around the neighbourhood and operate CCTV cameras which have been installed in the streets which fall under the auspices of Ekhaya (see Map 4.2 and figures below) – i.e. streets where social housing and private companies have purchased buildings and provided funds for their installation.

Bad Boyz, like other private security firms in South Africa (see Diphoorn, 2015a), deploy a tough, intimidating image. With authority, the company's founder, a former policeman who left the public service to start his own company, declares, 'We're not bad, but we don't tolerate crime, grime, any kind of misbehaving.' Studies of CCTV surveillance and regulating public space have described how authorities tend to operate on prohibitive or exclusionary logics and create social order by policing behaviours which are deemed to be harmful or undesirable (Coleman and Sim, 2000; Hayward, 2004; Koskela, 2000; Norris et al., 1998). Activities which challenge the commodified usage of space – particularly begging and sleeping rough, but also busking, skateboarding and loitering (Flusty, 2002; Fyfe, 2004; Langegger and Koester, 2016; Németh, 2006) – are increasingly being restricted and policed out of public urban spaces across the globe. As privatisation, gentrification, surveillance and commodification become the defining features of urban space, some scholars argue, the regularised, predictable

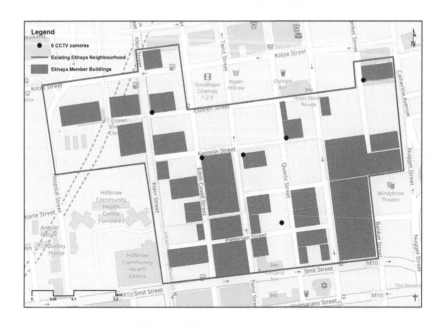

Map 4.2 Map of Ekhaya Neighbourhood.

and consumption-driven logic of the shopping mall becomes an almost ubiquitous form of, or substitute for, public life (Crawford, 1992). Indeed, this has come to pass in South Africa, where beggars, informal traders and homeless populations have become the targets of public and private police in CIDs and have often been forced out of these areas.

As the above quote from the company's founder demonstrates, Bad Boyz are engaging in a similar process, focusing their cameras and policing activities on the ways people act in and use the public spaces of the Ekhaya Neighbourhood. An employee of the company makes this clearer, as he explains that the guards monitoring the CCTV cameras focus on 'the movement and dressing' of people, and use these as indications of suspicious intent. In doing so, they are not only defining what potential criminals look like, but are also establishing the bounds of what constitutes acceptable or desirable ways of occupying space. He thus further explains that someone who walks with a specific, purposeful direction – i.e. a commuter – is regarded as acceptable, but someone who loiters or walks up and down a street repeatedly or circles around a block several times draws their attention. In these cases, private security personnel become the arbiters of what constitutes valid, purposeful behaviour in public space (cf. Cooper-Knock, 2016). Bad Boyz also define belonging on grounds of propriety and assumed class status, and use style of dress as indicators. In explaining how they utilise their CCTV

Figure 4.1 Surveillance camera in Hillbrow. Photograph by Thembani Mkhize.

Figure 4.2 Surveillance camera in Hillbrow. Photograph by Thembani Mkhize.

cameras and what signs guards are trained to look out for, the company's founder describes how

> somebody who goes to work, he'll have a tog bag, he'll have leather shoes, his shoes will be clean; the youngsters that's loitering [and consequently singled out as potential criminals] will be in groups of two or three, they'll wear hoodies, they'll have All Star takkies [trainers].

Thus, through the ways in which they regulate spaces, the company, on behalf of the investors who have hired them, are engaging in a form of social ordering, and are establishing the boundaries of acceptability and inclusion/exclusion in the area. They are, therefore, deploying and exercising spatial capital and producing spaces accordingly. The goal is the creation of an orderly, liveable neighbourhood, with a stable, employed residential population. It is, therefore, ultimately a project which aims to maximise the commercial viability and value of residential properties and ensure that investments are protected and profitable. As the coordinator of the Ekhaya Neighbourhood, who is employed full-time by the RCID and takes responsibility for organising public events, facilitating meetings between the different members and liaising with the municipality, relates, 'The main aim [of Ekhaya] is to help each other improve the area so that the business people in this area, the landlords and business owners, all those who have invested in Hillbrow, get more profit.' Security and policing play important roles, as they are tools which help shape the residential population and ensure that 'desirable' tenants are protected, and those who are deemed threating to commercial interests are policed and removed from the neighbourhood.

In addition to those who walk suspiciously and wear hoodies, the targets of surveillance and policing are homeless children living on the streets of Hillbrow and seemingly unemployed young men. Children living on the streets come in for particularly harsh attention. A housing supervisor illustrates the antagonistic relationship residents have with them. He further points out that part of the urban regeneration and management process has involved eliminating their presence, as they fall outside the bounds of the respectable, desirable community:

> Street kids, most of the people, they don't trust them because of this thing of stealing. You can leave them here and then they can steal your phone, they steal whatever it takes, they deceive. So that's a problem ... But they are not many in Hillbrow anymore, especially in our area ... we are working very hard to make the street to be clear and nice.

Furthermore, whilst the City's by-laws do not make gambling on the streets illegal, Bad Boyz focus their surveillance cameras on groups of young men who congregate on street corners and often play dice games. During the course of our interview, the Bad Boyz employee showed me a variety of still images of young men captured by the cameras. He points out that 'these are faces that we feel like they need to be monitored, people who need to be monitored because they are gaming [gambling] on the streets.' Because they do not have discernible employment or forms of income, they are assumed to engage in other deviant and criminal activities. Thus, he further explains, 'tomorrow if this person is in *The Star* [newspaper] because he's committed a crime, we can help the cops [locate him].' It therefore becomes apparent that a 'selective gaze' (Coleman, 2004; Coleman and Sim, 2000) is turned on the bodily habits and recreation activities of particular groups and that these come to be the signifiers of people's social identities and the bases on which they are judged to be either normal, respectable members of the community or threatening non-members. In an area with high unemployment rates, a large informal sector and a diverse population, the normative standard is established as a person who is employed, dressed professionally and visible. Those who do not meet these requirements threaten the identity of the community and the forms of behaviour which are deemed to be acceptable, and are consequently targeted and frequently removed. In these cases, fears of revanchism certainly are justified, and the neighbourhood is being constructed and shaped in ways which bear some hallmarks of the pacification and cleansing processes which have been witnessed in other cities (Mendes, 2014; Saborio, 2013; Schinkel and van den Berg, 2011; Swanson, 2007).

Everyday policing and social–spatial order

However, whilst revanchist forms of policing are indeed extremely disturbing, and cause real harm for those who fall victim to their violence, these forms of policing are not necessarily the norm, even in fractious, violent settings such as Hillbrow. They are certainly part of the policing and security landscape, but there are also more subtle, routine practices at work. In some ways, the academic literature has paid too much attention to the spectacular forms of policing, such as crackdowns on informal traders, and in doing so has mistaken these episodic practices for the standard form of policing. In doing so, this work has come to neglect or overlook the more mundane, everyday forms of policing which are constitutive of the social–spatial order (Cooper-Knock, 2016). The prevailing urban studies literature also fails to come to terms with the specific social and cultural logics which shape these practices, and therefore tends to overemphasise certain aspects, at the expense of fuller and more complex accounts.

Beyond the blitz approach

One Saturday night in 2014 I was accompanying the Hillbrow Community Policing Forum (CPF) on their patrol of the neighbourhood. Comprising a group of civilian volunteers and a police reservist, the street patrollers assembled at the local police headquarters and, after donning their orange high-visibility vests and going through a military-style roll-call, dispersed to canvas the streets of one of the busiest, most densely populated areas of the city. It is also an area notorious for its high levels of crime, violence and drug dealing. As the patrollers pointed out to me, they were on the look-out for petty muggers and people selling drugs. Unarmed, the patrollers set off in groups of two, with one overly enthusiastic young man breaking protocol and running out ahead of the others. I walked alongside the reservist, a tall, self-assured and commanding man who has been involved in crime fighting initiatives in the area for several years. Not long into our patrol, we turned into Quartz Street, a short stretch of road with several taverns and clubs crammed alongside one another. In due course we approached a young man exiting one of the taverns. Politely but authoritatively, the patroller introduced himself and asked if he could search the young man who, judging by his accent, was of Francophone African origin. Signalling that this was an interaction with well-rehearsed and familiar rules, the young man first demanded to see the patroller's hands to make sure he wasn't holding any drugs or weapons which could be planted on him, and then, satisfied that none were present, consented to being searched. After a vigorous frisking, which did not turn up any incriminating evidence, the patroller apologised to the young man and explained that it wasn't personal, he was just trying to keep the community safe (a refrain widely repeated by the patrollers). Shrugging his shoulders and with no apparent animosity or even anger at being subject to such an intrusive, physical search in public, the man simply declared, 'It's allowed,' and walked off into the night.

This moment resonated with me, and left me wondering how these intrusive, routine procedures have come to be accepted and part-and-parcel of everyday life in what was once the most notorious and violent area of the city. Whilst the use of force and coercion certainly are part of the story – the young man in the incident narrated above had little alternative but to submit to the patroller's demands – there are more complex, subtle processes at work too, which combine to make these procedures and forms of policing legitimate. For Murray (2008, 2011), crackdowns and the spread of private security forces in the inner-city are concomitant with the types of revanchist gentrification seen in cities around the world, and bear out the claim that the urban present and future is one based on homogenisation, elitism and anti-poor politics. According to McMichael (2015), 'blitzes' (swift, violent police raids on specific areas) represent the state's low-level, persistent war against urban disorder and exemplify the way state power is used to further the needs of the capitalist system. Whilst highlighting some crucial issues, these accounts collapse the state into one, monolithic entity, with one, monolithic mode of governance and agenda to fulfil. This is an over-simplification of what, by now and without reminding, should be recognised as a complex and varied social terrain. 'The State' can never be simplified into one single entity, pursuing a single objective (Gupta, 2013; Myers, 2011; Parnell and Robinson, 2012), even when it comes to policing (see Hornberger, 2004). As previous chapters have argued, whilst the post-apartheid state has been at the forefront of the roll out of neoliberal mechanisms and policies, it has also simultaneously, and in some cases successfully, pursued large-scale social redress and redistribution. Driven by and home to competing logics and practices, none of which can be said to prevail, 'the state' is a disorderly and contradictory entity, which often undermines its own operations (Chipkin, 2013). Therefore, rather than the modus operandi or *raison d'être* of the state and its way of governing, crackdowns and revanchist practices need to be understood as a particular instance and form of intervention, which draw on and are formed by particular logics and practices. Just as it is vital to draw attention to and resist the state's violent excesses, it is also imperative to recognise and analyse the other practices which are constitutive of urban social order.

Policing cultures and ideas about what constitutes legitimate policing are vital for the successful exercise of authority and the acquiescence of those who are being policed (Garland, 1996). In South Africa, this legitimacy is frequently absent, as communities continue to harbour hostile attitudes towards the police. During apartheid the police were not a public protection service but an extension of the state's military apparatus which terrorised black communities and enforced oppressive racial legislation (Marks, 2005). Currently, despite efforts to rebuild relationships between the police and communities, hostility continues to reign, which is exacerbated by corruption and the view that the police are predators, not public servants (Hornberger, 2004; Steinberg, 2008). For these reasons, Steinberg (2008) argues that policing in contemporary South Africa is a performance in which local communities, when it suits them and when police do not cross the boundaries of what they deem to be acceptable, pretend to consent to being policed and the police pretend to be policing. However, this pretence can give way at any point in time, and the police's ability to exercise order over public spaces is always tenuous and dependant on communities' moods. Given the disorder which generally prevails and which quickly returns to inner-city areas, blitzes and crackdowns need to be seen within this spectrum, as being 'staged performances' (Murray, 2011, p. 153) of state and policing power designed to give the illusion that order is being imposed and local government is paying attention to an area which frequently falls off the governance map (Winkler, 2008). The high profile and visibility of these campaigns, which are frequently announced in the media before being rolled out, is thus a strategy designed to prey on public sentiment and appeal to constituencies who are generally not located in the inner-city and continue to regard it as a zone of decay and despair. In a country still emerging from authoritarian rule and where violence is commonplace, it is a performance with considerable public appeal and which aims to respond to people's understandings of what government does and what governance looks like.

Similar forms of performativity are at work in the stop-and-search procedures employed by the CPF and the tough, zero tolerance image cultivated by Bad Boyz. The black population in South Africa have, generally, only experienced harsh, heavy-handed approaches to policing. Thus, stop-and-search procedures, once they have been carefully negotiated, are tolerated by the local population, as they are seen as the normal way in which policing functions (Vigneswaran, 2014). Furthermore, not only are they 'allowed,' they also respond to culturally ingrained norms and expectations of safety and effective policing. The influence of cultural perceptions over safety and policing is well established (for example see Brown and Benedict, 2002; Weitzer and Tuch, 2005). Thus, the tactics and approaches which are taken up by public officials and private service providers are those which are deemed most appropriate and, particularly in an era where security is one amongst other commodities to be purchased, desirable (Loader, 1999). Again, this helps to explain the discourse of blitzes and crackdowns which pervades in South Africa, as it responds to and reproduces the language and performance of authority that people feel constitute effective urban govern-ance and management.

Interviews with residents living in Hillbrow reinforced the desire for and, consequent legitimacy of, harsh, physical approaches to policing, including stop-and-search measurers. For example, when asked what would improve the neighbourhood, one young man declares,

> If they can put cameras around and the police are patrolling 24/7 you know that, *ai* [speaks forcefully and claps hands for emphasis], when you turn around this corner [clap], *there's a police, when you turn this side* [clap] *there's a police*, then you're safe there!

> (Tenant Three, Greatermans)

Another tenant expresses similar sentiments when, responding to the same question, she declares, 'if police are patrolling the streets, searching everyone who they suspect and making sure the kids cross the streets safely' (Tenant Five, Cavendish Court). In the context of the inner-city, legitimate policing is thus seen as a practice which is constant, indiscriminate, affected through daily routines and by policing people's bodies. Additionally, black communities in South Africa have long-standing practices of relying on community networks to police their neighbourhoods. During apartheid, because the police force terrorised, rather than protected communities, day-to-day order was enforced by street committees and vigilante groups, which continue to be influential in some areas (Buur and Jensen, 2004; Fourchard, 2011; Super, 2015). There is therefore a prevailing situation in which patrols, stop-and-search procedures and visible policing are all normalised and integrated into, rather than imposed onto, the social fabric of the neighbourhood, and come to reflect 'not only popular responses to vacuums left by state collapse and neo-liberalism, but also specific historical and cultural logics' (Fourchard, 2011, p. 611).

It is these logics and historical practices which allow harsh, intrusive policing to be regarded as part and parcel of everyday safety. Hence, stopping and searching people is part of the same continuum of everyday policing and regulation as helping children cross the streets. The street patrollers form part of this constellation and come to enjoy legitimacy because they enact the routines and practices which people understand as policing, and to do so on a rigorous, interpersonal scale. As one patroller proudly declares, 'No one can pass without us searching them. When they [the police] drive with the cars they pass criminals, but we are hands-on, we are searching almost everyone that we meet on the road.' He further points out that they feel they have widespread support amongst the community, which they are members of: 'People, I can say they like what we are doing because they are coming and joining us every now and then. We are getting new members almost every January and now we have got 170 or so.'

Earning legitimacy through spatial practice

The security regime which has come to the fore in Hillbrow also earns its legitimacy by fulfilling needs which go beyond basic safety. Bad Boyz use their surveillance cameras to keep track of the maintenance backlogs in the area, compiling records of blocked drains, damaged pavements, broken street lights and uncollected rubbish, which are reported to the City Council and relevant service providers. They also participate in cleaning campaigns around the neighbourhood, as seen in figures 4.3 and 4.4 below, and have even helped deliver babies when emergency services have failed to respond in time.[3] They therefore move beyond simply policing the population and protecting private interests and participate in providing the everyday repair, maintenance and care which an overburdened and often ineffective local state has been unable to deliver. As the company's founder points out, 'We're not actually a security

Figure 4.3 Bad Boyz street cleaning and maintenance.

Figure 4.4 Bad Boyz street cleaning and maintenance.

company, we're more an urban management company; we liaise with all the different departments, the cleaning companies, the Metro Police, all that kind of thing.' Similarly, the employee that I interviewed explains the outlook of the company and their practices as follows: 'This is a neighbourhood, we don't just look for crime only. We try always to promote the cleanliness of the neighbourhood, the improvement of the neighbourhood – that is urban management.'

Although their primary mandate is to protect private interests, Bad Boyz also come to serve the needs of residents, who require a clean and safe environment in which to live. In doing so, they also fulfil a developmental mandate of sorts, as they try and insist that maintenance and cleanliness standards in the inner-city are commensurate with those in the wealthier suburbs. As the company's founder declares,

> We will not tolerate them [the municipality] not giving the same service that the white people get in Sandton and Bedfordview and we get less service

here in Hillbrow. That's always my two areas that I measure service delivery: that *tannie* [old lady] in Bedfordview [a wealthy suburb in the eastern section of Johannesburg] won't take nonsense, that lady in Sandton won't take nonsense; why must we accept less?!

Thus, the private security regime in the inner-city has come to provide for more inclusive goals, which have enhanced the liveability of the area and aided in the formation of a community in the space. Private security has a notorious reputation for serving narrow, exclusive interests; in South Africa's predominantly white suburbs this is certainly the case. As private security patrols have been deployed and neighbourhoods fenced off, exclusive enclaves have arisen. Private security firms patrolling these exclusive spaces target members of the public who are deemed to not belong in the order which residents have paid them to protect and poor black people are regularly harassed and policed out of these areas (Diphoorn, 2017; Lemanski, 2006; Schuermans and Spoctor, 2016). Furthermore, these practices have created unequal landscapes of safety, where affluent suburbs enjoy higher levels of security and crime is pushed out to poorer neighbourhoods (Clarno, 2013; Clarno and Murray, 2013; Dirsuweit, 2007; Dirsuweit and Wafer, 2006; Landman, 2006). In Hillbrow, where private investment and redevelopment sit alongside urban dilapidation and decay, actors have been careful to avoid a similar situation. One of the main organisers of the CPF points out that they have cultivated good relationships with the private companies operating in the neighbourhood, and that a collective approach to securing the area prevails. He explains that they constantly engage with private security actors and property owners at public meetings, such as the Community Safety Forum, and remind them to be aware that 'security does not only end at your doorstep, but around [the area] and the people around the neighbourhood.' He further notes that 'you know that when you see any security company you can stop that person, you can stop that car and say, "Please help here," which they do.' Therefore, by serving a broader public, the private company begins to stray into territory ostensibly occupied by public police. Rather than focusing on the narrow security needs of clients, they play a significant role as a quasi-public service ensuring 'the maintenance of communal order, security and peace' (Baker, 2009, p. 9), which earns them legitimacy in the emerging broader social order. As a tenant living in social housing in Hillbrow points out, 'Most of the things they [Bad Boyz] do are for our protection' (Tenant One, Lake Success).

This case illustrates the role privatised services can play in shaping a broader social order, and how in the violent, fraught South African context, surveillance and policing are part of a broader continuum of social ordering and producing space. Non-state policing actors have increasingly come to perform security functions associated with the state (Davis, 2010; Diphoorn, 2015b; Jaffe, 2013). In South Africa, where the state is relied upon as a provider of basic services, this does not only apply to security personnel engaging in policing practices, but sees them taking part in other forms of service provision, social care and public management too. Moving away from a focus on policing as an episodic display of force, which the public police still tend to rely on, they come to involve themselves in much more routine activities. Private security services, then, are not only coercive and spectacular displays of force serving private interests (although the ability to mobilise violence remains part of the spectrum), but are part of the mundane activities which constitute and sustain everyday life in the inner-city.

Adaptive urban management

The security regime which has come into place has also been effective in the inner-city as it is not attempting to impose a form of social order onto people and the area, but works with and adapts to the realities of the space. The previous chapters have shown that the particularities and dynamics of the inner-city are key factors which shape housing providers' habitus and outlooks; the same is true for actors involved in urban management. Bad Boyz make concerted efforts to employ people who reside in Hillbrow and embed themselves in the community. The head of the company explains their approach to policing the neighbourhood as follows:

> The difference is everyone else, they outsource people from different areas and they come and dump them here by the buildings and they don't understand the concept or the environment that Hillbrow's created of. It's Nigerians, it's Zimbabweans, it's South Africans, it's Angolans, it's Russians, it's a mix of all races. So we've got a strict selection out of the area, we take people from the area that understand the concept, that understand the people, that understand the culture, they understand languages and we train them and leave them in the building and they adapt [to] the situation.

Adapting to the situation, as in the case of housing providers and finance companies, again becomes a central component of people's success in managing the area and acquiring spatial capital, which then allows them to effect change and contribute to regeneration. Reductions in the annual crime rates recorded for Hillbrow during the years in which I was conducting fieldwork illustrate the effectiveness of the policing strategies under consideration. For instance, there were 56 murders recorded in Hillbrow in 2012 and 63 in 2013, compared to 84 in 2008 and 88 in 2009.[4] Significant reductions are also recorded for attempted murder, assault with attempt to inflict grievous bodily harm, common robbery and robbery with aggravating circumstances between 2008 and 2013, pointing to the lasting effects of policing and security interventions in the area. Blitz campaigns and policing crackdowns have been mobilised in quests to reorder Johannesburg's inner-city in line with grand ambitious, such as securing World City status (Bremner, 2000; Gaule, 2005; Murray, 2008). They are thus externally imposed practices, which have only episodic effects on the area. It is widely accepted that after raids, informal traders will either return, or will be replaced by others engaging in the same activity (Kihato, 2007). In contrast, more durable and sustained policing practices entail a practical response that accepts the variety of needs and uses which the inner-city is now home to.

The head of Bad Boyz reflects this when, discussing informal traders, he states,

> The hawkers [informal traders], if they're off the street it will create that European feeling of the shops and all that thing, but we must accept this is Africa and accept this is a kind of single-man business and they also have the right to be here. They just need to be regulated.

Officially, informal trading is prohibited in the Ekhaya RCID, but practically officials adopt more tolerant, permissive attitudes. Trading remains widespread throughout the area, as figures 4.5, 4.6 and 4.7 illustrate, and a whole host of items is available, including clothes, cigarettes, roasted meat and mielies (corn), sweets, household goods and even live chickens. Mkhize (Forthcoming) shows that building managers, whilst officially saying the activity is prohibited, actually negotiate with traders and tolerate their presence directly outside their buildings, sometimes working with them to keep watch over public spaces and using them

Figure 4.5 Informal trading in Ekhaya. Photograph by Thembani Mkhize.

Figure 4.6 Informal trading in Ekhaya. Photograph by Thembani Mkhize.

to advertise vacancies, and even sheltering them from police raids on occasion. From his research, it becomes apparent that a form of mutuality and acceptance exists between traders and managers in the neighbourhood and urban management and policing are far more pragmatic and complex activities than they often appear to be. These pragmatic practices turn attention back to the agentful activities of those involved in urban regeneration and the ways in which they not only exercise spatial capital through dominating the space, but acquire capital by responding to the realities they are confronted with, and adapting their practices accordingly.

An Area Manager, who is employed by the largest for-profit housing company, which owns over 60 properties throughout the inner-city and thus has considerable power and capital (spatial and economic) in the area, demonstrates how the reality of the inner-city evades domination and forces actors to adapt to it; he illustrates how dealing with informal traders requires him to move away from the official requirements of urban management and regulation, and rather appreciate their circumstances in the wider social and spatial context instead. He explains that whilst he will not allow traders to work directly outside the buildings he is responsible for, he permits them to operate in close proximity and adopts a sympathetic attitude:

Figure 4.7 Informal trading in Ekhaya. Photograph by Thembani Mkhize.

You're kind of inclined to close a blind eye because remember, that's their livelihood, that puts food on the table for them on the end of the day [sic]. It's a 50–50 scenario; do I let them stay there and live or do I kick them off and tell them to get knotted? *And don't care?* Sometimes in life you need to close your eyes and accept what's coming your way, unfortunately. And this is Africa, *you need to.*

Although associations between 'Africa' and informality are simplistic, here he shows the way in which the space of the inner-city and the transformation it has undergone shapes his outlook on and practice of urban management. Rather than seeking to impose a form of social order, he, like others involved in urban management, acts through a habitus which adapts to the inner-city's circumstances and spatial realities (including being situated in Africa). Urban management in this case cannot be classified simply as interdiction (Flusty, 2002), revanchism (Smith, 1996) or 'taming' the inner-city (Murray, 2008). Rather, the process is characterised by an increased cognisance of the fact that cities are multiplicities and contain and are shaped by a variety of uses and inhabitants. The forms of urban management and regeneration which have emerged in the inner-city are not mimics of processes which have seemingly spread around the world, such as gentrification, revanchism, pacification and neoliberal urbanism. Rather, they are localised, hybridised adaptations to a varied social order which inculcates a variety of dispositions, outlooks and forms of action in people. Regenerating the inner-city thus entails multiple vernacular meanings, logics, practices and outcomes.

Domestication by football

The adaptive nature of the security and urban management practices which have developed in Hillbrow illustrate how a form of domestication has taken place. However, it is unlike the sort which has generally been documented and critiqued in the predominant urban studies literature to date. The term 'domestication by cappuccino' was popularised by Zukin (1995), as she sought to draw attention to the ways in which consumption practices are at the vanguard of gentrification procesess and class conflicts over urban space. She uses the phrase to point out how changing the character of a given area to suit upper-class tastes is a subtle way of colonising space and forcing others out. Domesticating spaces by making them more aesthetically appealing and welcoming to consumers also frequently presents a pleasant veneer to cover up the darker aspects of gentrification, including revanchist policing practices and the forceful removal of 'undesirables' (Atkinson, 2003). In Hillbrow, however, domestication is taking place in a different (although not always less violent or exclusionary) way.

Building communal relations

Transition and decline in Hillbrow did not only damage the physical landscape, but also left social relations in the area in a volatile, anomic state. What was once a relatively stable residential population was replaced by a transient, often furtive one, as single men, many of them undocumented migrants from other African countries, moved into the area in search of short-term accommodation (Morris, 1999a). Decayed buildings also became convenient shelters for criminals wanting to escape scrutiny and public visibility, and a general sense of fear, suspicion, animosity and anomie came to define the neighbourhood. Creating security and order in this context has therefore been a process of dealing with decayed buildings where possible, but also attempting to nurture and revitalise communal relations and ease a hostile social terrain. Urban regeneration has thus also come to focus on renewing or re-establishing a sense of communal belonging and collective solidarity in the area. Hence, the first coordinator of the Ekhaya Neighbourhood sums up the core activity of the project as liaising between different groups of people with interests in the area, 'getting people to know each other, reducing the space between people.'

Through the collective effort of housing companies, charities, local government and even Bad Boyz, a range of social and community-building initiatives have been started, under the blanket coordination of the Ekhaya Neighbourhood.

One particularly visible and popular project has been the creation of Ekhaya Park (shown in Figure 4.8 below). Located on the corner of Pietersen and Claim Streets, the park was built on a previously derelict lot which had been occupied by drug dealers and taxi drivers. Through a combination of forceful policing and communal initiative, the land was steadily reclaimed and converted into a temporary football pitch. This was not a seamless process, and drug dealers aggressively defended their territory, with one even attempting to ram his car into a crowd who had gathered for a makeshift football tournament. Although the full details of how the patch of ground was reclaimed remain slightly murky, eventually Ekhaya succeeded in securing the land and, through the efforts of the Johannesburg Development Agency (JDA) and funding secured from the 2010 Football World Cup legacy project, some jungle gyms and play equipment were installed and an astro-turf football field was eventually built on the site. At first glance it is easy to dismiss this as a cosmetic alteration. However, establishing the park stands as a significant achievement and addition to the neighbourhood for two main reasons. Firstly, the inner-city is densely populated, currently home to far more people than the area was originally designed to accommodate. It is a veritable concrete jungle, with few green spaces and recreation facilities. The situation has become even more acute as, as urban regeneration efforts have brought increased stability, household structures have changed in recent years, with areas like Hillbrow, Jeppestown and Berea becoming home to family-type living arrangements (Mosselson, 2017; see following chapters). Many people employed in the inner-city or nearby suburbs live with their young children, who have few places to play. It is therefore extremely welcome and important that a green recreation space was built and secured, right in the heart of Hillbrow.

Secondly, the park has played a central role in cultivating communal relationships between people. The football pitch has been central to this; several times each year teams comprised of children, and sometimes adults, from surrounding residential buildings come together to compete in tournaments. These were designed as vehicles through which community-building and friendships could be cultivated. By composing teams out of residential buildings, the organisers also encourage identification between people and the spaces in which they live, and help cultivate shared senses of belonging. As the current coordinator explains, 'We encourage them to play together so that they grow up knowing each other and knowing who is who.' In addition, Ekhaya also hosts events for children during school holiday periods, helps arrange life-skills training and workshops at the Hillbrow Theatre and also draws in adults who volunteer at the events in their spare time. Bad Boyz play a prominent role in these activities, donating equipment and making financial contributions, and also providing supervision and transporting the children from the various buildings to and from the park. Even on days when there are not organised events in Ekhaya Park, several security guards are present to keep watch over the children, as seen in Figure 4.9.

Through these innitiatives, important socialisation and recreation activities take place and help overcome some of what the current coordinator terms the

Figure 4.8 Ekhaya Park. Photograph by the author.

Figure 4.9 Security in Ekhaya Park. Photograph by Thembani Mkhize.

'anonymosity'[5] that contributed to the decay and violence which previously pervaded. As she reflects, in the early stages of regeneration in the neighbour-hood it was hard to organise residents or get people to interact: 'It was very difficult because you won't call anyone. For what? Everyone was minding his or her own business. No one wanted to know what's someone's business. Hence that anonymosity [sic] was what was promoting crime and everything.' The form of domestication underway, then, is not about attracting more affluent customers and taming the space to maximise the profit which can be extracted; rather it is a process which serves to create a neighbourhood in which a broader form of ordinary communal life can be staged, in what was once an unwelcoming and hostile urban environment (cf. Koch and Latham, 2013).

In addition to the park and the renovations which have been made to residential buildings, there have been several smaller interventions into the area's built environment. Unlike episodic blitz campaigns, ongoing maintenance is a key focus, and housing companies have pooled resources to pay for additional private cleaning services. Ekhaya also facilitates monthly meetings between housing companies' personnel and representatives of the City agencies responsible for refuse collection and public infrastructure maintenance. They therefore attempt to ensure that a managed, clean and orderly public environment is kept in place. Whilst these efforts are done with a view to maximising the attractiveness of the area, and therefore the value of property in it, they also respond to the needs of residents, who bear the brunt of poor management and have to contend with litter, blocked drains and broken pavements on a daily basis. They are also put in place as a way to produce order in the area, and are therefore part of the domestication and security efforts underway.

Atmospheres and ambient power in Hillbrow

Literature on the colonisation of public space emphasises how increased management, securitisation and alterations to the built environment are tactics which serve to introduce regulation, privatisation and consumerism into the public realm. The majority of this literature focuses on the criminalisation of certain behaviours (predominantly begging or pan-handling, but also skateboarding, busking and loitering) and shows how both disciplinary and coercive force are at work in these processes (Flusty, 2002; MacLeod, 2002; Németh, 2006; Smith, 1996). Whilst coercive force is used in the Hillbrow RCID at times, there are also more subtle processes at play. Recent work on regulation and surveillance has come to highlight the role of affect in creating atmospheres or settings, and how the feelings and dispositions engendered by particular assemblages of materials, technologies, architectural designs, and bodies shape how people experience and interact with particular social–spatial settings (Adey et al., 2013; Adey, 2014; Allen, 2006; Campbell, 2013; Hayward, 2004; Stewart, 2011).

A type of ambient, atmospheric control is in operation in Hillbrow. The intention hasn't (only) been to privatise the space or extract commercial value from it, but to shape the behaviours of people in it and make it a more manageable, orderly environment. When narrating the process of change in the neighbourhood, one example which stood out for interviewees was the cleaning up and closing of the alleyways between buildings (see Figures 4.10 and 4.11). During the 1990s and early 2000s, public services were almost non-existent in the neighbourhood. Refuse was not collected and simply left to pile up. Exacerbating the situation, the collapse of management structures in many buildings left residents to fend for themselves. Without any waste-disposal systems in place, people resorted to throwing garbage and sewerage from their apartments into the alleys between buildings. Even today, many remain clogged with litter, garbage and effluent. One of the first public infrastructure interventions was therefore to clean these spaces and seal them off so dumping could not take place. Reflecting on the process, the first coordinator of the Ekhaya RCID, who was instrumental in getting the regeneration process started, relates, 'The JDA fixed the lanes, then suddenly people stopped peeing [urinating] there; something triggered something.'

The 'something' she is referring to can be understood as 'ambient power' (Allen, 2006), the capacity of space to influence people's behaviours and the ways in which people read signals from the environments around them and shape their actions accordingly. In Hillbrow, people equate disorderly environments with crime and destructive social behaviours. One long-term resident of the area demonstrates this; when asked about littering and throwing objects from windows, and why he thinks people have now stopped, he exclaims:

> They were justified, they were justified! I didn't respect Hillbrow, I grew up in Hillbrow and I didn't respect Hillbrow. Like I said, there used to be tsotsis [criminals] here, now they are gone.
>
> (Tenant Three, Cavendish Court)

He argues that because crime rates were high and the area was in a state of disrepair, people were more inclined to engage in destructive behaviours. Conversely, as the area has improved and crime rates have declined, people have become more invested in the space and have adjusted their actions. By making incremental changes to the environment and improving the standards of maintenance and services, the managers in the RCID are able to shape people's conduct. The current coordinator points out that there has been a reduction in anti-social behaviour because, as people have come to see improvements in the built environment and cultivate stronger bonds with each other and the area, their attitudes have changed. As she explains, people 'are now being controlled by the area.' She argues that this is happening as people read signals from the environment around them. Thus, on entering the neighbourhood and seeing the new signs of order, upgrading and management, 'You will tell yourself this is not the right place for that [type of behaviour].' These interventions, then, have been used to shape people's actions and create a situation in which regulation is part of everyday life. It is a decidedly more subtle way of exercising power and shaping urban space, and rather than spectacular, coercive force, works with people's dispositions and attitudes. Whilst making it less visible and harder to identify, this subtlety also makes it more deeply ingrained in the area. One tenant living in Hillbrow sums up the changes and new feelings and experiences of the space which have arisen when she notes,

> Previously we were not feeling safe 'cause anytime someone can snatch your phone but now there's, like, cameras by the streets, cleaning staff, and there are security guards around the streets. I'm feeling nicer ... They are doing a good job.
>
> (Tenant One, Cavendish Court)

Security and communal life

Policing and claiming space

It therefore becomes clear that concerns about security and desires for management are not only forms of domination which serve the interests of property developers; they are also real concerns which affect the day-to-day lives of low-income residents living in the area, and others around the country (see Lemanski and Oldfield, 2009). When asked about improvements they would like to be made to the inner-city,

Figure 4.10 Closed alleyways in Hillbrow. Photograph by Thembani Mkhize.

Figure 4.11 Closed alleyways in Hillbrow. Photograph by Thembani Mkhize.

tenants living in social and for-profit housing frequently called for better and more consistent cleaning and maintenance, as the quotes below demonstrate:

> They must carry on upgrading the buildings. They haven't completed putting lights; there are still streets that are dark. There must also be an average number of people that can stay in an area. Maintenance is the biggest – everything else is here.
>
> (Tenant Two, Lake Success)

> They must deploy lots of cleaners, mustn't just have 20 people, must have some working morning, lunchtime and at night – three shifts! So it's cleaner. They must also monitor it – check if those people are doing their work. Maybe then it can be better. With crime it's already better.
>
> (Tenant Five, Lake Success)

Thus it can be seen that the desire for cleanliness is not only an upper-class habitus or reflection of World Class City aspirations, but is a concern of (low-income) people living in decayed urban areas too. This is summarised by a Bad Boyz employee, who argues, 'It's not good for our clients, it's good for anyone that will come here and pass through here that Hillbrow is clean and it's good.' By making neighbourhoods such as Hillbrow welcoming, safe and attractive (at least in comparison to its dirty, run-down state in the 1990s), urban management personal are enticing people into the area and encouraging them to use it in sociable ways. It is therefore as much a seductive experience of space (Allen, 2006) and use of spatial capital as it is one which relies on coercion and the capacity to use force (although this does still remain in place too).

Security patrols and policing in the area also serve a broader purpose. They enable a form of communal life to emerge and allow residents to participate in exercising spatial capital, and thus shaping the lived realities of the area themselves. When crime and violence were part of everyday life, a sense of powerlessness and detachment came to pervade amongst Hillbrow's residents. One building manager who arrived in the area in the late 1990s recounts:

> When I first came here to Johannesburg, sho, the muggings, the robbings, it was a daily thing; you'll get mugged during the day, broad daylight and nobody would do anything about it! Myself I never got mugged but I've seen a lot of people getting mugged and the tsotsis [criminals] were getting guns and knives and you would be afraid to confront them; it was horrible, actually.

In this context, policing crackdowns emerge as efforts to reclaim or reassert control over public space and return a sense of agency to local authorities. By the same token, everyday forms of policing serve a similar function, but empower a community, rather than a state or branch of local government, and are thus integral to processes of belonging, domestication and claiming space.

This is noticeable in ways residents narrate the current situation in the neighbourhood. Although fears of mugging remain, there was a shared sense that the area is much safer and the community, through both a private security company which serves its needs and collective mobilisations, now control it. Thus, one resident recounts, 'Fighting crime brought people together, when you see someone being attacked people try and help' (Tenant One, Lake Success). Another living in the same building paints a starker picture and reiterates how vigilante violence remains integral to community formation and policing in South Africa (see Buur and Jensen, 2004; Super, 2015, 2016). With some pride he declares,

> Hillbrow was very tough but [now] everything is cool. Now people have the authority to scream 'Help me!' and people will come. Before people were afraid to rescue you, now you get help easily. We beat whoever does crime and now we can move around at 23:00 or midnight and not be scared of anyone.
>
> (Tenant Three, Lake Success)

Another resident, originally from Zimbabwe, also draws attention to the ways in which enhanced feelings of safety have enabled people to feel greater senses of belonging in and ownership of public space. He points out, 'We've noticed heightened security, installation of cameras; you don't just do as you please on the illegal side of the law. You can almost walk tall!' (Tenant Six, Ridge Plaza). All of these quotes emphasise a sense of feeling empowered, demonstrating the relationship between belonging, community formation, policing practices and making/claiming space. The regime at work in Hillbrow has, in part, been effective because it has responded to a very real need people feel, and has given them a sense of renewed control over the area. The formation of 'community,' as vague and loose as this term may be, helps domesticate a space and make people feel at home in it. Similarly, exerting control over space helps facilitate the formation of particular forms of community. Thus the seductive power of policing comes to the fore, and the legitimacy and support afforded to street patrollers, regulatory regimes and Bad Boyz begins to makes sense, when they are understood in light of the localised logics (Lemanski and Oldfield, 2009) and relational, socially constructed desires and fears (Pow, 2015) which shape them and which they respond to.

Boundaries and exclusions

However, this is by no means to endorse or celebrate the practices at work in Hillbrow. Because a practice has local, vernacular logics and legitimacy, is widely supported and contributes to community formation does not mean that it is without problems or dangerous implications. The concept of 'community' is of course highly problematic and often serves as a stronger indicator of who/what is excluded, rather included (Staeheli, 2008). In Hillbrow, as much as the policing measures at work attempt to serve inclusive needs, they also serve as a form of social ordering in which boundaries of inclusion/exclusion are drawn.

I have already described how homeless people, particularly children, are targets for hostility and are constructed as threats to the community which has come into being. Residents also share this antagonism towards them, and continue a pattern in which perceived criminals are constructed as outside the boundaries of the acceptable South African citizenry and are consequently seen as legitimate targets for violence, antipathy, arrest and elimination (Super, 2016). As one tenant recounts,

> There were more than 30 street kids at the corner of Smit Street – they're no longer there, I don't know where they are. Someone saw they're a threat because you don't know what they are going to do. But at least I'm happy that they're gone.
>
> (Tenant Five, Lake Success)

In addition to street kids, the population living in derelict buildings is singled out as threatening and beyond inclusion in the 'good' community. In interviews with urban management personnel, local government representatives and tenants, hijacked or 'bad' buildings were frequently mentioned as obstacles to the regeneration process or problematic features of the inner-city. The terms 'building hijacking' or 'hijacked buildings' are widely used, but not well understood in Johannesburg. Broadly, they speak to processes whereby landlords lose control of their buildings, particularly their ability to collect rental income. In the early and mid-1990s there were instances of armed gangs taking over buildings using violence and forcing tenants to pay rent to them rather than the legitimate owners. In most cases, however, 'hijackings' took place when tenants resolved or were convinced to withhold rent from landlords and pay monthly fees into trust accounts set up by other parties who promised to look after maintenance on the tenants' behalves. However, many of these trusts were fraudulent or were taken over by criminal syndicates who pocketed the money and gained de facto control of the buildings. In other instances, they created conflicts over ownership and management, and led to many owners abandoning their properties and buildings falling behind on their payments to the City, and subsequently being disconnected from water and electricity services. However, not all derelict buildings are 'hijacked.' The majority are sectional title properties, in which each unit is owned by a different person. During the process of capital and residential flight, many owners abandoned their units, sold them informally or even gave them away (Morris, 1999a, 1999b). As they did so, collective management bodies within buildings collapsed, leading to deficits in maintenance and payment for services and buildings subsequently being disconnected from electricity and water and falling into ruin. In other cases, desperate people have forced their way into abandoned properties and established their own improvised living arrangements (Mayson and Charlton, 2015).

Whilst the conditions in derelict and hijacked buildings are similar, with no access to water and electricity, overcrowding, improvised living arrangements and insecurity and health and safety hazards (COHRE, 2005), there are significant differences between buildings which are occupied by people who cannot afford to live elsewhere in the inner-city and ones which have been taken over by criminals. In general discourse, however, no distinction is made. This is an act of what Bourdieu (1984) terms 'symbolic violence,' as it constructs people who would otherwise be homeless and who cannot afford the rents charged in formalised buildings as criminals and threats to the community and social order which has been established. It delegitimises their claims to spaces and housing and promotes a vision of the area in which all residents are paying customers and live in formalised housing. In this instance, it becomes clear how people associate particular types of space with particular behaviours and social categories. As one housing supervisor, laments,

> That situation [of derelict buildings], I feel very bad about it because that's a situation that builds, that grows criminality. Because in those kinds of buildings you don't have any kind of management, most of those buildings, when they rob people they run into those kinds of buildings, the criminals run into those buildings, and it creates criminals.

A local resident who participates in the Hillbrow CPF also conflates the state of a building with behaviours which threaten the community and includes these types of buildings amongst the factors that are contributing to the presence of street kids and crime in Hillbrow:

> There are abandoned buildings in Hillbrow where these kids, they even stay in these buildings … there's no owner of the building and they just stay there, that's where they hide. They are dangerous those buildings … there are plenty actually.

Although it is easy to see how these buildings do indeed make ideal places for criminals to hide, the practice of seeing them exclusively as dens of illegality ignores and obscures the important role they are actually playing in providing shelter for people who cannot afford the rentals being charged in formalised buildings or cannot meet the entry requirements demanded by housing companies. Symbolic violence is therefore not only perpetrated against the buildings themselves, but against the inner-city poor, who come to be criminalised through the ways in which they access housing. This practice continues a prevalent tendency to de-link those who commit crimes (or who are perceived to) from structural and political contexts, and focus antipathy or hostility on criminals instead (Super, 2016).

Furthermore, as much as security and management personnel adopt inclusive rhetoric and practices, they are also, ultimately, providing privatised services. Although there is no formal boundary drawn around the Ekhaya Neighbourhood and it is far more porous than other CIDs and enclosed suburbs around Johannesburg, the establishment of security remains contingent on investment and property companies having interests to protect. Whilst social cohesion in the area has improved dramatically, as with other instances of 'broken windows' approaches to policing (see Sampson and Raudenbush, 2004), the key drivers of crime in the inner-city – particularly unemployment, substance abuse and lack of care facilities for vulnerable populations – are not addressed, and crime is being displaced to surrounding areas which have not attracted private investment. Due to the self-avowed 'zero tolerance approach,' drug dealing has largely been pushed out of the RCID (although it does still take place in certain 'hot spots' which are the preoccupation of street patrollers and private security personnel (Vigneswaran, 2014)), and now concentrates in spaces a few blocks away. The head of Bad Boyz is very frank in admitting that crime has been displaced, rather than alleviated, but does not see this has his concern. Unapologetically, he reflects,

> The crime just moved on, it moved out to Yeoville, it moved out to Jeppe, Jewel Street and we've got a happy police force here [in the RCID] because it's much less for them to do and they're more accepted by the community now.

It is clear, then, that a differential form of policing is being put in place, which continues to concentrate security around areas which have attracted private investments and where powerful companies have interests to protect. Whilst the improvement district does not have visible boundaries around it, it is demarcated by the limits of the influence of property owners, and at present only has effect in areas in which there are enough people contributing to pay for private services. The northern section of Hillbrow currently does not benefit from street patrols, privatised cleaning and CCTV cameras as there are not enough property owners in that part of the suburb making financial contributions to pay for these. Plans are underway to expand the district, but these are contingent on companies owning enough property to be able to control the public spaces of the area. Thus, the regeneration approach, whilst achieving developmental results, is also compounding inequalities. Because it has been formulated by private companies, whether they are social housing or for-profit, and has not been part of a coherent strategy to secure the entire area and improve it, it is giving rise to a fragmented urban landscape. So, whilst the language of 'fortified enclaves' and 'privatised fiefdoms' that Murray (2008, 2011) adopts is overly alarmist and dramatic, there is also a drastic need for truly inclusive approaches to the social issues which drive crime and insecurity to be adopted. Additionally, the types of crime which private companies focus on are selective (Koskela, 2000). Whilst 'suspicious' activities like gambling and loitering attract attention, violent, invasive practices which target women's bodies escape scrutiny. Cat-calling and even inappropriate touching, are not picked up by personnel monitoring security cameras, and leave women feeling greater degrees of anxiety on the streets than most men. Again, the limitations of approaches to security which prioritise investment and private interests come to the fore, and remain areas of concern, particularly in a society with disturbing levels of gender-based violence.

Conclusion: between revanchism and the everyday

There are therefore new, revanchist forms of exclusion emerging in the inner-city at the same time as new community relations and forms of associational life are being established. Whilst they attempt to be inclusive, urban management and securitisation inevitably draw boundaries and come to benefit some people at the expense of others. This is apparent in the ways crime is being displaced from areas controlled by Bad Boyz, and children living on the streets, unemployed young men and the occupants of derelict buildings are being singled out as criminals. Furthermore, inequalities in service delivery and the safety and quality of the environment are emerging between the RCID and areas which have not yet attracted investment. It therefore becomes clear that the everyday practices which are making regeneration possible reflect and reproduce the duality of the wider regeneration process. Urban management strategies are furthering the commercial interests of housing companies, whilst simultaneously pursuing developmental goals and attempting to domesticate spaces in ways which are conducive to creating a shared sense of public life. However, just as the benefits of regeneration are not shared by all and the process is raising property and rental prices in the area and displacing some people, the benefits of security and belonging are limited to those who meet the normative standards determined by housing companies and management personnel. Both are thus part of a process of establishing a contradictory social order in the area which is based on neoliberal practices as well as transformative goals.

At the same time, just as housing providers are shown to adjust their outlooks and practices to accommodate the complexities and diversity of the inner-city, urban management and security are also shown to be contingent, adaptive, everyday processes. Whilst urban management and security personnel attempt to create social order and enforce standards of behaviour, they also come to accept practices and forms of urbanity which are less predictable and orderly. They therefore make some allowances for the diversity of the inner-city's population and their different needs and survival strategies. The ways in which regeneration is being carried out are again shown to reflect the local, spatial dynamics of the inner-city. Tolerant approaches also illustrate that regeneration is dualistic not only in the ways it pursues two agendas, but also in the ways in which it both produces but also adapts to spaces. It is this adaptive nature which has allowed management practices to be successful and gain support from the local community, as they regard them as serving their needs and responding to their desires and demands. Security and management strategies are thus not simply imposed onto people and the area, but reflect the experiences of local communities and their needs for safety, particularly in dangerous, stressed environments.

Private security and regulation are also not simply pacifying public spaces and bringing them under control; they are allowing people to make use of these spaces and form attachments to the area and each other, thereby encouraging rather than hindering social life. Existing work on private policing and the privatisation of public space frequently misses this duality or ambiguity, painting pictures instead of a relentless march towards the homogenisation of all public spaces everywhere. Whilst trends highlighted in this body of work are real and disturbing, the narratives outlined here demonstrate that security, safety and community formation, whilst inherently exclusionary, also make life viable for people, particularly in stressed and dangerous urban environments. If we are to take the experiences and realities of people living in these contexts seriously, then an appreciation of this ambiguity is vital.

Appreciating ambiguity and contradictory outcomes is also crucial for evaluating the regeneration process itself. The chapters which follow will further advance this argument and describe the everyday lives which are coming into being because of regeneration. They will explore tenants' experiences of living in renovated social and affordable housing developments and demonstrate the contradictory, multi-faceted effects the process is engendering. Firstly, Chapter 5 will demonstrate that the provision of centrally located housing is aiding in spatial integration and transforming apartheid geographies, but that tenants remain marginalised in the inner-city's social order and live in in-between states, working hard to maintain their positions in the area but not necessarily wanting to live there.

Secondly, Chapter 6 will examine the conditions inside residential buildings; it will become clear that these buildings enforce strict hierarchies which regulate tenants and protect property companies' commercial interests, even at the expense of rights and protections granted to tenants. However, it will also demonstrate how tenants' everyday lives and forms of habitation remake these spaces and contribute towards changing what the area represents, yet again underscoring the ambiguity and duality of the regeneration process.

Notes

1 www.jda.org.za/index.php/milestones/fashion-dist.
2 Following a case brought before the Constitutional Court, it has been determined that property owners cannot be compelled to pay additional levies, even if the majority of owners in an area agree in principal. Thus, all CIDs have been made informal and voluntary.
3 www.news24.com/SouthAfrica/News/guards-cleaners-deliver-baby-on-hillbrow-pavement-20151218.
4 www.crimestatssa.com/precinct.php?id=261. These figures were obtained from the website crimestats.sa.com, and verified against the statics made publicly available by the South African Police Service (SAPS). SAPS' figures were obtained from www.saps.gov.za/services/crimestats_archive.php.
5 Although she is potentially mis-speaking, I prefer to retain this phrase as an excellent indication of how anonymity and animosity were intertwined in the neighbourhood previously, and in the thinking of those involved in producing security and order in the area.

References

Adey, P., 2014. Security atmospheres or the crystallisation of worlds. *Environment and Planning: Society and Space.* 32, 834–851.
Adey, P., Brayer, L., Masson, D., Murphy, P., Simpson, P., Tixier, N., 2013. 'Pour votre tranquillité': ambiance, atmosphere, and surveillance. *Geoforum.* 49, 299–309.
Allen, J., 2006. Ambient power: Berlin's Potsdamer Platz and the seductive logic of public spaces. *Urban Studies.* 43, 441–455.
Atkinson, R., 2003. Domestication by cappuccino or a revenge on urban space? Control and empowerment in the management of public spaces. *Urban Studies.* 40, 1829–1843.
Baker, B., 2009. *Security in Post-Conflict Africa: The Role of Nonstate Policing.* CRC Press. Boca Raton, Florida.
Beall, J., Crankshaw, O., Parnell, S., 2002. *Uniting a Divided City: Governance and Social Exclusion in Johannesburg.* Earthscan. London.
Bénit-Gbaffou, C., 2009. In the shadow of 2010: democracy and displacement in the Greater Ellis Park Development project, in: Pillay, U., Tomlinson, R., Bass, O. (Eds.), *Development and Dreams: The Urban Legacy of the 2010 Football World Cup.* HSRC Press. Cape Town, pp. 200–222.
Bénit-Gbaffou, C., Dubresson, A., Fourchard, L., Ginisty, K., Jaglin, S., Olukoju, A., Owuor, S., Vivet, J., 2013. Exploring the role of party politics in the governance of African cities, in: Bekker, S., Fourchard, L. (Eds.), *Governing Cities in Africa: Politics and Policies.* HSRC Press. Cape Town, pp. 17–42.

Bourdieu, P., 1984. *Distinction: A Social Critique of the Judgement of Taste.* Harvard University Press. Cambridge, Massachusetts.

Bremner, L., 2000. Reinventing the Johannesburg inner city. *Cities.* 17, 185–193.

Brown, B., Benedict, W.R., 2002. Perceptions of the police: past findings, methodological issues, conceptual issues and policy implications. *Policing: An International Journal.* 25, 543–580.

Buur, L., Jensen, S., 2004. Introduction: vigilantism and the policing of everyday life in South Africa. *African Studies.* 63, 139–152.

Campbell, E., 2013. Transgression, affect and performance: choreographing a politics of urban space. *British Journal of Criminology.* 53, 18–40.

Chipkin, I., 2013. Whither the state? Corruption, institutions and state-building in South Africa. *Politikon.* 40, 211–231.

Clarno, A., 2013. Rescaling white space in post-apartheid Johannesburg. *Antipode.* 45, 1190–1212.

Clarno, A., Murray, M.J., 2013. Policing in Johannesburg after apartheid. *Social Dynamics.* 39, 210–227.

COHRE, 2005. *Any Room for the Poor? Forced Evictions in Johannesburg, South Africa.* Centre for Housing Rights and Evictions. Johannesburg.

Coleman, R., 2004. Watching the degenerate: street camera surveillance and urban regeneration. *Local Economics.* 19, 199–211.

Coleman, R., Sim, J., 2000. 'You'll never walk alone': CCTV surveillance, order and neo-liberal rule in Liverpool city centre. *British Journal of Sociology.* 51, 623–639.

Cook, I.R., 2010. Policing, partnerships, and profits: the operations of Business Improvement Districts and Town Center Management Schemes in England. *Urban Geography.* 31, 453–478.

Cooper-Knock, S.J., 2016. Behind closed gates: everyday policing in Durban, South Africa. *Africa.* 86, 98–121.

Crawford, M., 1992. The world in a shopping mall, in: Sorkin, M. (Ed.), *Variations on a Theme Park: The New American City and the End of Public Space.* Noonday Press. New York, pp. 3–30.

Davis, D.E., 2010. Irregular armed forces, shifting patterns of commitment, and fragmented sovereignty in the developing world. *Theory and Society.* 39, 397–413.

Didier, S., Peyroux, E., Morange, M., 2012. The spreading of the City Improvement District model in Johannesburg and Cape Town: urban regeneration and the neoliberal agenda in South Africa. *International Journal of Urban and Regional Research.* 36, 915–935.

Diphoorn, T., 2015a. 'It's all about the body': the bodily capital of armed response officers in South Africa. *Medical Anthropology.* 34, 336–352.

Diphoorn, T., 2015b. Twilight policing: private security practices in South Africa. *British Journal of Criminology.* 56(2), 313–331.

Diphoorn, T., 2017. The 'Bravo Mike Syndrome': private security culture and racial profiling in South Africa. *Policing and Society.* 27, 525–540.

Dirsuweit, T., 2007. Between ontological security and the right to difference: road closures, communitarianism and urban ethics in Johannesburg, South Africa. *Autrepart.* 42, 53–71.

Dirsuweit, T., Wafer, A., 2006. Scale, governance and the maintenance of privileged control: the case of road closures in Johannesburg's northern suburbs. *Urban Forum.* 17, 327–352.

Eick, V., 2012. The co-production of purified space: hybrid policing in German Business Improvement Districts. *European Urban and Regional Studies.* 19, 121–136.

Flusty, S., 2002. The banality of interdiction: surveillance, control and the displacement of diversity. *International Journal of Urban and Regional Research.* 25, 658–664.

Fourchard, L., 2011. The politics of mobilization for security in South African townships. *African Affairs*. 110, 607–627.

Fyfe, N., 2004. Zero tolerance, maximum surveillance? Deviance, difference and crime control in the late modern city, in: Lees, L. (Ed.), *The Emancipatory City? Paradoxes and Possibilities*. Sage Publications. London, pp. 40–56.

Garland, D., 1996. The limits of the sovereign state. *British Journal of Criminology*. 36, 445–471.

Gaule, S., 2005. Alternating currents of power: from colonial to post-apartheid spatial patterns in Newtown, Johannesburg. *Urban Studies*. 42, 2335–2361.

Gupta, A., 2013. Messy bureaucracies. *HAU Journal of Ethnographic Theory*. 3, 435–440.

Hayward, K., 2004. *City Limits: Crime, Consumer Culture and the Urban Experience*. Routledge. New York.

Hornberger, J., 2004. 'My police – your police': the informal privatisation of the police in the inner city of Johannesburg. *African Studies*. 63, 213–230.

Jaffe, R., 2013. The hybrid state: crime and citizenship in urban Jamaica. *American Ethnologist*. 40, 734–748.

Kihato, C., 2007. Governing the city? South Africa's struggle to deal with urban immigrants after apartheid. *African Identities*. 5, 261–278.

Koch, R., Latham, A., 2013. On the hard work of domesticating a public space. *Urban Studies*. 50, 6–21.

Koskela, H., 2000. 'The gaze without eyes': video-surveillance and the changing nature of urban space. *Progress in Human Geography*. 24, 243–265.

Landman, K., 2006. Privatising public space in post-apartheid South African cities through neighbourhood enclosures. *GeoJournal*. 66, 133–146.

Langegger, S., Koester, S., 2016. Dwelling without a home: Denver's splintered public spaces, in: De Backer, M., Melgaço, L., Varna, G., Menichelli, F. (Eds.), *Order and Conflict in Public Space*. Routledge. London, pp. 140–159.

Lemanski, C., 2006. Residential responses to fear (of crime plus) in two Cape Town suburbs: implications for the post-apartheid city. *Journal of International Development*. 18, 787–802.

Lemanski, C., Oldfield, S., 2009. The parallel claims of gated communities and land invasions in a Southern city: polarised state responses. *Environment and Planning A*. 41, 634–648.

Lippert, R., 2012. 'Clean and safe' passage: Business Improvement Districts, urban security modes, and knowledge brokers. *European Urban and Regional Studies*. 19, 167–180.

Loader, I., 1999. Consumer culture and the commodification of policing and security. *Sociology*. 33, 373–392.

MacLeod, G., 2002. From urban entrepreneurialism to a 'revanchist city'? On the spatial injustices of Glasgow's renaissance. *Antipode*. 34, 602–624.

Marks, M., 2005. *Transforming the Robocops*. UKZN Press. Scotsville, South Africa.

Mayson, S.S., Charlton, S., 2015. Accommodation and tenuous livelihoods in Johannesburg's inner city: the 'rooms' and 'spaces' typologies. *Urban Forum*. 26, 343–372.

McMichael, C., 2015. Urban pacification and 'blitzes' in contemporary Johannesburg. *Antipode*. 47, 1261–1278.

Mendes, A.F., 2014. Between shocks and finance: pacification and the integration of the favela into the city in Rio de Janeiro. *South Atlantic Quarterly*. 113, 866–873.

Miraftab, F., 2007. Governing post apartheid spatiality: implementing City Improvement Districts in Cape Town. *Antipode*. 39, 602–626.

Mkhize, T., Forthcoming. *Urban Crime and Grime: Lessons from Hillbrow's eKhaya Residential City Improvement District, Urban Innovations. Studying and Documenting*

Innovative Approaches to Urban Pressures. Department of Planning Monitoring and Evaluation.

Morris, A., 1999a. *Bleakness & Light: Inner-City Transition in Hillbrow, Johannesburg.* Wits University Press. Johannesburg.

Morris, A., 1999b. Tenant–landlord relations, the anti-apartheid struggle and physical decline in Hillbrow, an inner-city neighbourhood in Johannesburg. *Urban Studies.* 36, 509–526.

Mosselson, A., 2017. 'It's not a place I like, but I can live with it': ambiguous experiences of living in state-subsidised rental housing in inner-city Johannesburg. *Transformation: Critical Perspectives on Southern Africa.* 93, 142–169.

Murray, M.J., 2008. *Taming the Disorderly City: The Spatial Landscape of Johannesburg After Apartheid.* Cornell University Press. Ithaca, New York.

Murray, M.J., 2011. *City of Extremes: The Spatial Politics of Johannesburg.* Duke University Press. Durham, North Carolina.

Myers, G.A., 2011. *African Cities: Alternative Visions of Urban Theory and Practice.* Zed Books. London.

Németh, J., 2006. Conflict, exclusion, relocation: skateboarding and public space. *Journal of Urban Design.* 11, 297–318.

Norris, C., Moran, J., Armstrong, G., 1998. *Surveillance, Closed Circuit Television, and Social Control.* Ashgate. Farnham.

Paasche, T.F., Yarwood, R., Sidaway, J.D., 2014. Territorial tactics: the socio-spatial significance of private policing strategies in Cape Town. *Urban Studies.* 51, 1559–1575.

Parnell, S., Robinson, J., 2012. (Re)theorizing cities from the Global South: looking beyond neoliberalism. *Urban Geography.* 33, 593–617.

Peyroux, E., 2012. Legitimating Business Improvement Districts in Johannesburg: a discursive perspective on urban regeneration and policy transfer. *European Urban and Regional Studies.* 19, 181–194.

Pow, C.P., 2015. Urban dystopia and epistemologies of hope. *Progress in Human Geography.* 39, 464–485.

Saborio, S., 2013. The pacification of the favelas: mega events, global competitiveness, and the neutralization of marginality. *Socialist Studies.* 9, 130–145.

Sampson, R.J., Raudenbush, S.W., 2004. Seeing disorder: neighborhood stigma and the social construction of 'broken windows.' *Social Psychology Quarterly.* 67, 319–342.

Schinkel, W., van Den Berg, M., 2011. City of exception: the Dutch revanchist city and the urban homo sacer. *Antipode.* 43, 1911–1938.

Schuermans, N., Spoctor, M., 2016. Avoiding encounters with poverty: aesthetics, politics and economics in a privaleged neighbourhood of Cape Town, in: De Backer, M., Melgaço, L., Varna, G., Menichelli, F. (Eds.), *Order and Conflict in Public Space.* Routledge. London, pp. 79–100.

Smith, N., 1996. *The New Urban Frontier: Gentrification and the Revanchist City.* Routledge. New York.

Staeheli, L.A., 2008. Citizenship and the problem of community. *Political Geography.* 27, 5–21.

Steinberg, J., 2008. *Thin Blue: The Unwritten Rules of Policing South Africa.* Jonathan Ball with Open Society Foundation for South Africa. Johannesburg.

Stewart, K., 2011. Atmospheric attunements. *Environment and Planning: Society and Space.* 29, 445–453.

Super, G., 2015. Violence and democracy in Khayelitsha: governing crime through the 'Community.' *Stability: International Journal of Security and Development.* 4, 1–20.

Super, G., 2016. Volatile sovereignty: governing crime through the community in Khayelit-sha. *Law and Society Revue.* 50, 450–483.

Swanson, K., 2007. Revanchist urbanism heads south: the regulation of indigenous beggars and street vendors in Ecuador. *Antipode.* 39, 708–728.

Vigneswaran, D., 2014. The contours of disorder: crime maps and territorial policing in South Africa. *Environment and Planning: Society and Space.* 32, 91–107.

Viruly, F., Bertoldi, A., Booth, K., Gardner, D., Hague, K., 2010. *Analysis of the Impact of the JDA's Area-Based Regeneration Projects on Private Sector Investments: An Overview.* Johannesburg Development Agency. Johannesburg.

Weitzer, R., Tuch, S.A., 2005. Determinants of public satisfaction with the police. *Policing Quarterly.* 8, 279–297.

Winkler, T., 2008. Reimagining inner-city regeneration in Hillbrow, Johannesburg: identifying a role for faith-based community development, in: Harper, T.L., Gar-On Yeh, A., Costa, H. (Eds.), *Dialogues in Urban and Regional Planning.* Routledge. New York, pp. 133–149.

Zukin, S., 1995. *The Cultures of Cities.* Wiley. Hoboken, New Jersey.

5 Ambiguous experiences of regeneration

Spatial capital, agency and living in-between

Introduction

This chapter takes a closer look at the effects the regeneration process is having on lived-experiences of the inner-city, and seeks to understand what the process has meant for tenants living in renovated buildings. It demonstrates that the contradictory nature of the regeneration process is lived out on an everyday basis; the ways in which people experience and inhabit residential buildings and make use of the city bear the hallmarks of both the market-driven, restrictive aspects of the process as well as some of the more transformative and developmental sides of it too. Framing the discussion through the lens of spatial capital, it becomes clear that residents are, on one level, experiencing improved forms or stocks of spatial capital, particularly as this relates to abilities to live centrally and enjoy the benefits of access to urban amenities. However, as earlier chapters have discussed, spatial capital also entails abilities to engage in making and shaping spaces, and is a reflection of social hierarchies and power relations. As the discussion in this chapter unfolds, it will become apparent that residents in renovated social and affordable housing developments are in many ways marginalised, both in terms of the day-to-day running and management of the buildings in which they reside, as well as in the broader spatial order of the inner-city. Thus, whilst the regeneration process is going some way to enhancing basic rights and needs, it is also limited and is certainly not an effort to upend or alter pre-existing power differentials and patterns of marginalisation. The experiences documented here are therefore illustrative of the ways in which transformation of the South African social order is restricted and partial; whilst significant changes have taken place in the post-apartheid era, severe inequalities, limitations and stratifications remain in place.

Integration and urban change

Enhancing access to urban amenities, improving spatial capital

In the context of Johannesburg's enduring geographies of fragmentation, inequality and segregation, integrating people into central areas is a significant step towards realising urban citizenship and improving the spatial capital which low-to-moderate income households can lay claim to. Previous chapters have outlined how the current regeneration process fits into government's ambitions for densification, creating more inclusive cities and giving poor communities better access to urban amenities and opportunities. Several interviewees confirm that these goals are, in some instances, being met. Access to transport and work opportunities are highlighted as crucial benefits people gain by residing in the area. As one tenant explains,

> I'm close to anything – transport, everything; it's closer to work, I walk to work so for me it's much better.
>
> (Tenant Two, Greatermans)

Many other tenants repeat this narrative, and emphasise how they are able to save money, despite the relatively high costs of accommodation. For example, another points out 'In Thembisa [a township situated in Johannesburg's East Rand] we spend too much on transport but now we can save, expenses are gone. We are saving a lot' (Tenant Two, Cavendish Court). Several younger interviewees moved to the inner-city to study and 19 out of the 57 tenants interviewed explicitly stated they chose to reside in the area because of the nearby employment opportunities. Many of those in the interview sample who were in formal employment (some interviewees were studying, some were not working because their partners did) were employed in jobs located in or around the inner-city, including working at government offices or banks or as security guards or domestic workers in the near-by suburbs.

It thus becomes clear that access to housing in the inner-city is aiding in changing residential patterns, improving spatial integration and thus helping to realise some of the promises of the post-apartheid period, for some households at least. Opportunities to live centrally, to access jobs, services and to make something of the city are crucial elements of urban citizenship (Huchzermeyer, 2011; Simone, 2008). They are not end points in realising the right to the city, but vital initial steps towards achieving this ideal (Parnell and Pieterse, 2010); the provision of social and affordable housing in the inner-city should be thought of as part of this process. The way in which it assists in developing spatial capital is also highly significant. If spatial capital is taken to encompass the abilities of households to ensure or protect their social reproduction, then the ability to live in close proximity to places of work and education becomes invaluable indeed.

In addition to facing financial hardships, people living on the peripheries of South African cities also tend to be 'time poor' and spend many hours commuting, often in inefficient, stressful and at times even dangerous conditions (Kihato, 2014, p. 366; Mokonyama and Mubiwa, 2014). By being able to live in an area that provides them either easy access to transport networks or lets them live in close proximity to places of employment, tenants gain opportunities to save money, which can then be spent on improving their lifestyles, saving or investing or educating their children (thus helping secure the social reproduction of households). They also gain better quality of life and can spend more time raising their children, relaxing or caring for themselves and others. As one tenant explains, now that she is living in the central part of the city, her commuting times are drastically reduced, and she is consequently able to be a better parent to her children:

> Let's say I get out here at 7 o'clock, and then I will reach MTN [i.e. Noord Street taxi rank] say five-past seven; then I will be able to get in Midrand at 8 o'clock. Rather than if I'm staying in Soweto, because sometimes then I have to use the bus, and the bus is very slow. I have to wake up at 4 o'clock. Maybe at 4 I have to be on the road, although I'm going to get to work at 8 o'clock; that's very inconvenient for me because I woke up early and I get in the house late again. I'm going to get in the house maybe 8 o'clock, and then I've got kids then when I get here I'm tired, I can't play with them, even if it's a short time. The only thing that I need is to go to bed. That's the thing that I can say, at least in time consuming, at least it's [living in the inner-city] better.
>
> (Tenant One, Cavendish Court)

Rérat and Lees (2011) include ease of mobility as an important component of spatial capital. This is the case not only because of time savings and access to employment opportunities but also because it fosters abilities to live in one place and work in another. In the European context, this ability is crucial for upwardly mobile, transnational professionals, and helps secure their privileged positions in the economic hierarchy. In the South African context, the spatial capital and ease of mobility that are gained by living centrally are crucial for working-class households' daily survival. The bulk of enterprises and employment opportunities are located in the northern sections of the city-region (Gotz and Todes, 2014). However, accommodation in these areas is also prohibitively expensive. Therefore, many people move to the inner-city as it is more affordable and better connected than far-flung townships or suburbs. Affordability of accommodation also means that people are able to live with their families, and therefore reproduce their households, which is another important component of spatial capital. Through the improvements which have been made to individual buildings and the wider urban environment, the section of Hillbrow which has been targeted by regeneration initiatives has become more stable, safer and attractive for family-type living arrangements. As the Ekhaya coordinator points out, the success of the regeneration process is evident in the fact that the area has become 'family friendly':

> Ekhaya has been successful because it has now become the home of where people live. The working people live here – schoolchildren with their families, actually. Families can now live in the Ekhaya buildings, not like before; before you'd never live with your family in Hillbrow. It was a place of someone who's working and [families] are at home. It was never stable

but now Ekhaya has made Hillbrow to be the stable home for people who live in it.

This is a drastic change from both the apartheid period, in which black households were broken up by influx control laws, the migrant labour system and forced removals, as well as the period when the area was in a severe state of decay. Enhanced urban management and policing and tight regulations in residential buildings, whilst having exclusionary effects (as discussed in detail in the subsequent chapter), have contributed to changing the social fabric of the area, and are helping some low-to-moderate-income households realise aspects of their urban citizenship. These developments are evident in the way the Johannesburg Housing Company's tenant population has changed over time. In 2006, 46% of their tenants were single people, either living alone or sharing units. However, in 2012 only 21% of their units were occupied by single adults, whilst 34% consisted of nuclear families and a further 33% were occupied by couples who did not have or live with children (JHC, n.d.). The research sample my study is based on also reflects the latter pattern as, out of the 57 tenants interviewed, 42 were living with their families, whilst only 10 were living alone or with people they were not related to. Thus, if urban citizenship and spatial capital are taken to include good or decent quality of life and the ability to nurture one's social relationships and devote time to the things that matter and make life enjoyable (Sen, 2001), then it becomes clear that the provision of affordable and social housing in the inner-city is indeed contributing to creating new and better experiences of the post-apartheid city.

Improving social environments, but obstacles to change

The changes which have taken place in terms of tenants' households have also led to changes in the social environments inside buildings. Housing providers, although utilising strict management practices and often harsh measures to enforce rental collection (as will be detailed later) make concerted efforts to cater for the new family-type living arrangements in their buildings. Social housing companies provide extensive tenant support services, including social work and counselling services, extra-curricular programmes and supervision for children. Some private companies have also invested in educational and child-care facilities in their buildings or the wider area. As one private developer describes,

> You'll find that there's a lot of children in the buildings, there's many families that are intact which is exciting, there's more family cohesion and social cohesion now and it does create challenges for us in terms of facilities for the children where the buildings typically didn't have playground facilities or other entertainment avenues for the children ... We do have two crèches that we sponsored in our bigger buildings; half of the intake will be from the building and half from the surroundings.

Improvements such as these, although not widespread in the inner-city, have added important comforts and support for some tenants and are allowing them to establish more settled, stable and comfortable ways of being in the inner-city.

However, these demographic changes have also been uneven, brought new challenges and exposed some of the enduring inequalities of the post-apartheid city. Many tenants living with children complain about the poor conditions which still prevail in the area and the lack of spaces for recreation. As the following tenants point out:

> There is no space for them. This place is not a good place for the children, not at all.
>
> (Tenant One, Cavendish Court)

> For my kids, shame, they've got no place to play, especially this one [younger child], he's got no place to play. He would love to play soccer if there's a small space, but they've got nothing, it's so difficult for them.
>
> (Tenant Five, Greatermans)

> To grow up in Hillbrow is not good.
>
> (Tenant Four, Lake Success)

As the developer quoted above noted, many of the residential buildings which have been renovated are also not well equipped for family living. Interviewees frequently made mention of the dangers the lack of recreation spaces in high-rise residential buildings pose for children. One incident, in which a child fell to her death inside a building, stands out as a particularly jarring and sad case which several interviewees made mention of. Additionally, whilst rentals are comparatively affordable, they do remain high, and mean that tenants have to improvise their living arrangements. In many instances families share one-bedroom apartments, with the parents sleeping in the bedroom and the children sleeping in the living room. In other cases, entire families share one room, utilising cabinets, bookshelves or even curtains as room dividers. These tactics, although allowing families to stay together, introduce tensions and lead to a lack of privacy and personal space. There are also issues in communal housing, where tenants complain about having to share facilities with families with children, citing noise and mess as problems.

As the school-going population in the area has increased, the severe shortage of education facilities available has also been exposed. This highlights the lack of coherent planning that has accompanied the regeneration process, as well as the limitations of relying on a market-driven and property-based approach to urban renewal. By viewing the inner-city as a source of revenue and almost exclusively as an investment destination, authorities have failed to take account of the social needs of residents living in the area, and have subordinated integrated social and spatial planning to the drive to increase revenue streams. Whilst the income derived from the inner-city is indeed vital for the city's maintenance, this neglect aggravates the hardships tenants living in the area have to endure and brings the inequalities which continue to define Johannesburg's spatial landscape into stark relief. Thus, when asked whether the inner-city is a good place to raise children, one tenant bitterly and emphatically states,

> No, absolutely no! Not even in Berea. Town [i.e. the CBD] and Hillbrow, they're no sort of place to raise children. You raise them because there's nothing you can do but it's not a good place to raise children.
>
> (Tenant Two, Ridge Plaza)

She also draws attention to the disparities between wealthy, formerly-white suburban areas and the inner-city, which are still obviously evident, and exclaims,

> You know I don't think there's a child who's better off, from here up to town. I think the children from Sandton, Fourways, those suburbs there, those ones are getting a better life or maybe a better environment.
>
> (ibid.)

Thus, it is apparent that whilst positive changes have been made to the urban environment and some households' living conditions, the inequalities of the wider city continue to shape people's experiences, identities, claims to citizenship and place in the spatial order. The regeneration process, then, continues to be a stark, physical expression of the limitations and unevenness of post-apartheid society and transformation continues to represent a number of disparities and ambiguities.

Between the township and the suburb

Temporary geographies

In addition to the challenges faced by parents trying to raise children, other living conditions in the inner-city also continue to highlight disparities and contradictions in the way the post-apartheid social order has evolved. For the most part, tenants regard living in the inner-city as an improvement, particularly in comparison to life in the townships. For instance, a woman working as a nurse in a retirement home in the north-eastern suburbs explains that she chose to move to the CBD because of the ease and comfort it gives her:

> It's easy for me to get taxis and the shops are around me. If I knock off late at work I just take one taxi, instead of two or three to get to the lokshin ... In the case of transport, all the taxis end here. It's very easy here in Jozi [the colloquial nickname given to the city]. I don't have stress.
>
> (Tenant Four, Cavendish Court)

A young woman living in a for-profit development in the CBD sums up the differences between the inner-city and the townships and gives a strong indication of the enjoyment tenants derive from the improved standards and accessibility of the inner-city: 'Like at home, when you want to buy groceries you have to catch two taxis, and it's dusty. Here you can walk freely, and the shops are near, the schools are near; everything is number one here!' (Tenant Five, Cavendish Court). Another, when asked if living in the inner-city provides her with opportunities and experiences which she did not have when she was living elsewhere exclaims,

> A lot, a lot, a lot! When I need something right now I can just go for it. When I'm at the location I have to think about it, I have to go with the transport so that's a very bad time than when I'm here.
>
> (Tenant One, Rochester)

These comparisons draw attention to the ways in which the townships for many still represent places of lack or hardship, and how living in them continues to cut people off from urban amenities and opportunities for social mobility and infuses daily life with difficulties. The townships' ongoing association with deprivation and apartheid-era forms of segregation is illustrated by tenants' retention of the word 'location' or 'lokshin,' terms used during colonial and apartheid times to refer to areas which were designated as black urban settlements. Whilst abject depictions of the townships have rightly been challenged and alternative narratives have been provided (see Dlamini, 2009), the tenants involved in this study persisted in painting them as unpleasant, impoverished and remote places. The significance of urban regeneration which enhances the liveability of the inner-city and the accessibility of affordable housing in the area is thus further emphasised and the ways in which tenants living in the area come to enjoy new forms of spatial capital and improved experiences of urban citizenship too are brought to attention. It also becomes clear how the regeneration process is creating opportunities for new settlement and residence patterns and is thus playing a vital role in undoing apartheid-era geographies.

However, these experiences remain ambiguous and caught within a sharp contradiction. Whilst the inner-city is regarded as a step-up from the townships, many tenants continue to experience it as a difficult, harsh and unpleasant place. Despite the positive changes which have been made, the majority of tenants interviewed were not enthusiastic about living in the inner-city, and, as other studies have found, regard it as a temporary destination (Kihato, 2013; Landau, 2006; Winkler, 2008). Out of the interviews which yielded information about tenants' long-term plans, 15 expressed a desire or planned to stay in the inner-city for the long-term, whilst 28 did not regard it as a permanent residential option.[1] Many tenants continue to think of other areas as 'home' and continue to orientate their social and cultural lives elsewhere. For instance, a young couple living in a for-profit building in the CBD continue to think of Soweto and the outer suburbs of Pretoria, where they are originally from respectively, as the places where they feel at home and spend most of their weekends outside of the inner-city (Tenant One and Two, Greatermans). A resident in a for-profit building in Berea declares that 'Home is Limpopo' (Tenant Three, Ridge Plaza). Another tenant aspires to return to Ladysmith, in Kwa-Zulu Natal, when it is her time to retire and longingly states, 'The only thing I think about is going home' (Tenant Six, Rochester).

Accepting one's place

Tenants' decisions to move to the area, and consequent relationships with it, are results of the difficult quests to secure livelihoods and reflections of the ongoing spatial fragmentation and inequalities which characterise contemporary Johannesburg. When asked why they chose to move to the inner-city, tenants frequently cited either the lack of other affordable options in better parts of the city or the ongoing difficulties living in the townships present. As one tenant explains, 'Hillbrow is the only place, the only solution. If you stay in the location the money spent on rent and transport is equal to what I pay here' (Tenant Three, Lake Success). Another tenant, when asked if Hillbrow is a good place in which to raise children laughs and says simply, 'I don't have a choice' (Tenant Five, Gaelic Mansions). Putting it more starkly, another states, 'I'm not staying in Hillbrow because I like drugs and crime – it's affordability' (Tenant Nine, Lake Success). Furthermore, another explains that he came to live in Hillbrow because 'it was the cheapest place I could find, I can't go to Sandton or Morningside [upmarket suburbs in the northern region of Johannesburg]' (Tenent Seven, Lake Success).

As the above quotes illustrate, the majority of tenants draw geographies for themselves which divide Johannesburg into three parts: the poor and under-developed townships, the expensive and beyond-reach formerly white suburbs, and the unpleasant but accessible inner-city.[2] On the one hand, this again emphasises the importance of affordable inner-city accommodation and reaffirms the positive development that more housing options are being created in the area. On the other hand, vast disparities remain in place and regenerating the inner-city does little to change the city's broader patterns of inequality. The harsh and difficult realities of Johannesburg's inner-city mean that tenants have to endure conditions which are far from ideal and have a tendency to regard themselves as stuck in a place they do not want to be in. Despite improvements, the inner-city remains troubled by physical decay, poor levels of service and maintenance, overcrowding, inadequate social amenities and high levels of crime. Tenants living in the area therefore must constantly negotiate these conditions and adjust their everyday realities to suit them. Negotiations, trade-offs and compromises are part of urban life for people in other parts of the city, or cities everywhere, too. However, the difficulties people face take on a particular, vernacular form in the inner-city. The negotiations tenants have to make signify how the inner-city represents upward mobility and improved urban access, but is also a place of violence and vulnerability which makes Johannesburg's disparities a lived, everyday reality.

For example, one tenant relates that she was witness to extreme violence and that this has affected how she navigates the city and makes use of its public spaces:

> I saw a person being shot back there at Tudhope [one of the main streets in Berea] so ... It's difficult but you have to learn to deal with it and take care of yourself, you need to put yourself in order; certain time such things they make you stay in the house.
>
> (Tenant Three, Ridge Plaza)

Another (somewhat hyperbolically) also emphasises how proximity to violence is an all-too-real part of life for people in the inner-city, but that it is something they are willing to endure due to the locational advantages they accrue: 'It's not safe here, people get killed every day. But it's nice, you can get everything for cheap, it's near everywhere, schools and taxi ranks' (Tenant One, Cavendish Court). Tenants also complain about the litter, traffic, noise and overcrowding in the area, and thus continually harbour ambitions to leave. However, because of Johannesburg's geographies of severe inequality, they remain trapped, not wanting to return to marginal townships and unable to find alternative affordable accommodation in better areas. Lefebvre (1991) and scholars inspired by his work argue that positions in physical space reflect as well as reproduce people's positions in the socio-economic order (Prigge, 2008; Stanek, 2008, 2011). In the case of inner-city residents, their place comes to signify the in-between nature of capital and citizenship in the post-apartheid era for many working-class or low-to-moderate-income black households. Oldfield and Greyling (2015) argue that poor communities who are waiting for formal housing to be delivered by the state live in conditions of permanent temporariness. They construe this condition as both disabling and agentive, as citizens are reliant on a largely indifferent and opaque state, but simultaneously make 'quiet encroachments' which allow them to counter state inactivity and secure footholds in the urban space economy (Oldfield and Greyling, 2015, p. 1100). Similarly, tenants in inner-city Johannesburg live in conditions of temporariness and inhabit what Kihato (2013) terms 'in-between' geographies, spaces and conditions in which they are aware that their lives are improving in some ways, but continue to aspire to move on, but frequently lack the means to do so.

Tenants absorb these in-between geographies and spatial experiences into their identities and habitus, and thus in-betweenness becomes a social and spatial expression of their position in the social order and the limited possibilities which it gives rise to. For Oldfield and Greyling (2015, p. 1102), 'waiting for housing is a process that, in mundane and profound, short-term and life-long registers, shapes what it demands and means to claim rights and citizenship after apartheid.' Something similar can be said about tenants living in the inner-city and the ways in which endurance and prolonged temporariness shape their dispositions, lived realities and experiences of the post-apartheid city, and thus their citizenship too. Whilst the bulk of tenants aspire to leave the area, they come to accept that for the foreseeable future they will be unable to, and thus must inure themselves to coping with these harsh realities and adjust their aspirations, identities and habitus accordingly. As one tenant living in social housing reflects, 'For now I'm stuck here, but I don't have a problem' (Tenant Seven, Lake Success). Another demonstrates how recognition of Johannesburg's spatial inequalities and their precarious position within the city's social order means that tenants are grateful for what they can find; pragmatically he points out, 'We'd like a garden, but beggars can't be choosers' (Tenant Eighteen, Gaelic Mansions). Further indicating the way tenants recognise their conditions, and consequently adjust to them, another explains, 'It's just that you have to accept the condition, the way of living where you are. You have to accept' (Tenant Two, Ridge Plaza), whilst another sums up the way many relate to the inner-city by stating pragmatically, 'It's not a place I like, but I can live with it' (Tenant Six, Rochester).

On agency and acceptance

The agency of everyday endurance

Habitus and spatial capital are not only reflections of positions in the socio-economic hierarchy; they are also concepts which help draw attention to the ways in which hierarchy is reproduced through lived, spatial experiences and actions (Bourdieu, 2005; Gale, 2005; Hillier and Rooksby, 2005; Holt, 2008). Furthermore, agency does not only exist in moments of resistance or when people challenge existing social structures. Rather, it is a mundane, everyday thing. Compliance, quiet endurance or reproducing existing structures also require choice and action. Thus tenants, although finding themselves in difficult, restricted and temporary geographies, should not be regarded as lacking agency. The very fact that they choose to live in the inner-city over other (albeit limited) options is a confirmation that they are not simply passive or prone victims of circumstance. Being socially distant from others is also often a choice people make, as they prefer the anonymity urban life offers over complicated and sometimes restrictive communal and personal obligations (Kihato, 2014; Landau, 2018; Thrift, 2005). The ways in which tenants continue to endure and make lives for themselves in the inner-city, despite the difficulties and unpleasantness this may entail, are thus also expressions of agency. What needs to be recognised is that agency is a reflection of habitus, or that habitus and social conditions influence the directions and activities through which agency is exercised (Bourdieu, 1990). Giving the concept an overtly spatial perspective draws attention towards the way in which experiences in space also shape the course through and objectives to which agency is exerted. In these ways, agency, although never absent, is always a socio-spatial outcome, and a means through which spatial and social orders are expressed and replicated on a day-to-day basis.

This point is brought home by the fact that, although they do not necessarily want to remain in the inner-city and direct their aspirations elsewhere, tenants work hard to continue to afford the rents they have to pay, and devote significant amounts of emotional and physical energy to maintaining their positions within the area. Whilst the accommodation being provided is partially state-subsidised, the commercial imperatives of housing companies and escalating costs of basic services mean that rentals remain relatively expensive. Although some are able to save money, others complain about the expenses they accrue. One tenant in for-profit accommodation points out, 'The main problem that we are having is the rent issue, it's affecting us big time; it's putting a strain on our lives' (Tenant One, Ridge Plaza, Berea). Another tenant also exclaims, 'I'm paying too much, the money is too much! The rates are too high here. Times are expensive' (Tenant Three, Constantine). These problems are also shared by social housing tenants, as this interviewee indicates: 'Every month people move out because of [the costs of] rent and electricity' (Tenant Two, Lake Success).

In addition to creating stress and forcing some people to move, these financial burdens also have effects on their experiences of the city and dispositions towards its spaces. They do not take away tenants' agency, but shape the course to which it is exercised. Various scholars have demonstrated how marginalised positions in the socio-economic hierarchy translate into people's conceptions of themselves and their abilities to speak about or act on a range of issues. Numerous studies point to the ways in which young people experience education institutions differently, depending on their geographic location, class positions and associated habitus, and adjust their aspirations and career plans in accordance with the social environments and worlds they grow up in (for example see Allen and Hollingworth, 2013; Appadurai, 2001; Bhat and Rather, 2013; Bourdieu and Passeron, 1990; Gale and Parker, 2015; Zipin et al., 2015). Similarly, gentrification research which mobilises the concepts spatial capital and habitus demonstrates how upper-class professionals feel entitled to claim and shape urban environments, and expect to occupy spaces which reflect their tastes and preferences (Benson and Jackson, 2013; Butler and Robson, 2003; Centner, 2008). In Johannesburg's inner-city, when the relationship between habitus, space, aspiration and disposition comes into focus, it becomes apparent how people's marginal economic positions result in them having precarious footholds in the urban space economy, which in turn shape their capacities for acting and the ways they engage with the city. Long histories of dispossession, marginality, struggles to get by and everyday endurance shape what people expect and feel entitled to, and consequently how they appreciate and relate to the spaces in which they live.

Habitus and limited spatial capital

Because tenants are predominantly concerned with their economic survival and getting by in stressful circumstances, their abilities or desires to engage actively with the spaces in which they live are limited. Many complained about and were frustrated by the lack of social amenities and recreation facilities and were unaware of or unable to participate in the social activities offered by social housing companies or Ekhaya. Consequently, most tenants' energy and agency is directed towards working and finding ways to get by, and instances of community engagement either focus on crime-combatting initiatives, or are organised by dominant actors, such as Ekhaya. As one tenant bitterly complains, 'It's like you are working only for paying the rent. What next about life [sic]? Nothing you can do' (Tenant Three, Ridge Plaza). Another also complains, 'There's no space here, there's nothing, just for your own living, just to come sleep, eat and go to school, that's the space that we have here' (Tenant One, Cavendish Court).

Thus, many tenants experience the inner-city as a place where financial concerns, limited means, difficult living conditions and endurance prevail. Although Ekhaya and other non-governmental and faith-based organisations do offer moments of collective life and enjoyment, and tenants' lives are not without instances of friendship, hope, solidarities and forms of resilience (as will be shown later), their experiences are largely characterised by feelings of detachment and resignation, and their agency is shaped by having to endure a harsh, far-from-ideal living environment. Instances of active citizenship or communal organisation are few and far between (Malcomess and Wilhelm-Solomon, 2016). Therefore, if spatial capital is taken as encompassing abilities to take and make place, it becomes apparent that tenants are lacking in this power for a number of reasons: firstly, they direct their energies and agency to ensuring economic survival, and thus lack the time, motivation and means to engage in making places and asserting their positions in the inner-city landscape. As one social housing tenant bitterly reflects, 'People in Hillbrow are working, going to church, doing night shifts – they do not have the time to protest' (Tenant Nine, Lake Success). Secondly, because many residents regard the area as a temporary destination, people are not invested in it or concerned with expending energy mobilising or trying to improve it. Many simply prefer to concentrate on working and concerning themselves with personal issues and struggles. A local councillor sums the situation up as follows:

> Here in inner-city it's not easy of that [sic] [to form community and become politically engaged] because most of the people that are staying in the inner-city are from different provinces, they only influx here for a better living, that's the only thing they are here for. They are not interested in anything, so if they can get a job and work, after that, by the end of the month they just go home.

A housing supervisor also echoes this when, when asked if he thinks that there is a communal atmosphere in Hillbrow he responds,

> I wouldn't say that much, because everybody here is in Joburg looking for a job. It's not like when you are coming from in the township or in the rural

areas where it's in a community and we grew up together, because every-body here comes from a different place, so everybody is looking for a job, looking for their own living to survive.

Thirdly, the economic necessities which shape their lives tend to become absorbed into people's identities and habitus, and consequently define the scope of their imaginations, aspirations and concerns to which they direct their agency. Thus, mobilising against crime becomes one of the focal points of communal action, as this is the biggest concern many tenants have, and is also an area of life they are able to have some control over, particularly given the high levels of policing and securitisation in the area. When it comes to larger questions and spatial imaginations, however, many remain detached and indifferent. Gevisser gives a poignant account of this when he narrates the difficulties researchers encountered when trying to include Hillbrow residents in the design of Constitutional Hill. He recounts that residents who were interviewed were not even aware of the construction taking place right on their doorstep, as their fears of detection due to their immigration statuses, involvement in covert economic activities and stressful lives meant that residents did not engage with or pay attention to what was happening around them, and instead focused their attention on getting by and surviving in the day-to-day (Gevisser, 2008, 2014). Here he gives a powerful account of the intertwined relationship between identity, positions in the social hierarchy and abilities to engage with and actively shape space.

Although the tenants I interviewed were in less precarious situations that the people Gevisser discusses, the scale and scope of the problems in the inner-city tend to make them equally inclined to feel passive and unable to affect change in the environment. When tenants were asked about possible improvements that could be made to the inner-city, there were very few suggestions offered which placed community and active citizenship at the centre. The quote below stands out as one exception:

> I think if we can try to form a community or something that we can always come together as people, sit down, come up with different ideas how to improve as a group, 'cause as an individual it's not easy to get improve [sic] but as a group you can make something big and better.
>
> (Tenant Eight Gaelic Mansions)

Rather, the majority, when choosing to proffer suggestions,[3] adopted passive stances, using the vague third-person pronoun 'they.' Replies such as these were characteristic and frequent:

> I don't know. If they could clean the area it would be fine.
>
> (Tenant Five, Rochester)

> There are other buildings that are old. If they can renew them it would be better.
>
> (Tenant Two Gaelic Mansions)

It is much safer but if they could make it more safer it would be better.

(Tenant One, Lake Success)

There are still some buildings which are neglected. In those there is no law and order, it is too much crowded. If they could modify them so they look nice [that would improve the area].

(Tenant Four, Lake Success)

Thus, tenants generally do not see themselves as active members of the inner-city community and misrecognise their abilities to actively make and alter spaces. This situation reflects and is exacerbated by the unequal distribution of political and spatial capital in the area. Housing companies and investors are the dominant political forces and actors and they monopolise governance procedures and processes, often at the expense or exclusion of residents. Their economic and spatial capital allow them to alter environments, including streetscapes, security arrangements and entire buildings in ways which tenants are unable to. Tenants are aware of these disparities, reflected in the way they surrender their creative capacities to 'them.' It thus becomes clear how positions in space and the forms of capital which different groups have at their disposal reflect as well as reproduce positions in the political hierarchy and order of the area. Bourdieu (1990, p. 131) asserts that habitus implies 'a sense of one's place, but also 'a sense of the place of others.' Tenants are content to surrender responsibility and abilities to effect change to housing companies, security and management personnel and local government, feeling that it is not their place to be involved in these activities. At the same time, as they do so, they also resign themselves to living in the area in passive ways, and do not think about individual and communal capacities to mobilise and shape space, even through small, incremental steps. Communal struggles, forms of protest, self-help and social support are prevalent in many settings in South Africa (Alexander, 2010; Desai, 2002), but are largely absent from the lives of inner-city residents. Urban citizenships is thus incomplete and severely restrained in the post-apartheid period.

Conclusion

The changes brought about by the regeneration process are thus palpably uneven and partial. There is no doubt that important improvements have been made to sections of the inner-city, and that these have positive influences on tenants' lives. The ways in which people gain access to amenities, employment opportunities, time and space to raise their families and improved urban lifestyles certainly have to be factored into any assessments of the regeneration process. In light of these, it needs to be appreciated as making a significant contribution to improving some aspects of people's rights to the city, urban citizenship and the forms of spatial capital they enjoy. However, these changes cannot be over-romanticised or regarded too optimistically. Life in the inner-city remains difficult and brings with it a range of pressures, stresses, trade-offs and negotiations. It is thus a product and reflection of the limited scope of change and the 'progress, setbacks and stagnation' which characterise the post-apartheid period (Kihato, 2014, p. 357).

Although some instances of densification and improved urban integration are occurring, these remain limited and, as this and other chapters have pointed out, city-wide change has been and continues to be difficult, if not impossible, to achieve. Instead, people make the most of limited possibilities and experience transformation in an uneven and unequal way. Geographies of inequality and exclusion remain defining features of the South African spatial landscape, and tenants' experiences, and the types of worldviews, dispositions and expressions which these give rise to, are reflective of this. They are also powerful reflections and reminders of the ways in which physical space is intrinsically linked to social space and positions in the social hierarchy. Thus tenants, although happy to claim chances for social mobility and to access better living conditions, simultaneously have to forgo other rights (as the following chapter will detail) and learn to live with harsh, sometimes dangerous, situations. In these ways, it becomes clear that change in South Africa is ambiguous, partial and contains a multitude of experiences and possibilities, ranging from submission to domination, resistance, stubborn as well as quiet endurance, community mobilisation, chosen or enforced social distance and estrangement, hope, aspiration and resignation. Temptations to resolve its meanings into one experience or term – neoliberalism, colonial domination, democratic empowerment – need to be avoided and an appreciation of the complex, ambiguous and vernacular nature of what the post-apartheid moment represents needs to be maintained and further drawn out.

Although part of a particular vernacular, there is also no reason for the insights derived from tenants' experiences of living in inner-city Johannesburg to be confined to the case at hand. Building a cosmopolitan and plural urban studies discipline and practice means theorising from a variety of contexts and extrapolating ideas, concepts and theoretical frameworks from multiple locations. Thus, experiences from Johannesburg can and should speak back to and inform how we think about experiences in other places too. Just as ideas gleaned from Bourdieu and Lefebvre can 'travel' and help frame how we understand events that are unfolding in Johannesburg's inner-city, a world very far removed from the conditions these scholars were working in and theorising from, so too can the insights derived from this specific context be used to think about and understand events and processes elsewhere. Thus, this chapter has aimed to offer ways of thinking about experiences of change, the relationship between agency and space and the conditions which shape and constrain active urban citizenship which can be of use to thinking through similar issues in a variety of contexts. The chapter has sought to highlight the complex, always-negotiated and spatially mediated nature of agency, and to draw attention to the fact that agency and choice are not only visible or realised in moments of resistance or non-conformity; those who struggle to make the best of difficult conditions, who find ways to endure and wait out hardships, and even those who partake in quiet acquiescence are making choices, adapting to and living out socio-spatial realities and inequalities. For instance, when it comes to gentrification, it is important to not only study the ways in which affected communities mobilise and try to resist processes of displacement, but to also turn attention to the everyday ways in which people negotiate processes of spatial change, attempt to endure and secure places for themselves in neighbourhoods undergoing gentrification, build stocks of capital and social networks which allow them to adapt to change, and are even conflicted by it. Turning our attention to these acts does not make our work less political, but, I would argue, rather makes it more so, as it allows us to deal with the conflictual, complex and indeterminate nature of individual and collective identities and experiences, and thus helps us arrive at better understandings of social reality. It also affords people more respect and as it acknowledges their multiple positionalities and sometimes conflicting, ambiguous reactions.

Notes

1 Information regarding this question was not obtained from 14 interviews.
2 There are, of course, other residential areas and options across the city. What is important here is that these descriptions and geographies are being drawn by the tenants being interviewed. They are therefore not presented as factual descriptions of the city, but as indicators of the subjective experiences and ways in which tenants have constructed the post-apartheid city.
3 Frequently, tenants were unable to make any suggestions and simply shrugged or laughed off the question.

References

Alexander, P., 2010. Rebellion of the poor: South Africa's service delivery protests – a preliminary analysis. *Revue of African Political Economy*. 37, 25–40.

Allen, K., Hollingworth, S., 2013. 'Sticky subjects' or 'cosmopolitan creatives'? Social class, place and urban young people's aspirations for work in the knowledge economy. *Urban Studies*. 50, 499–517.

Appadurai, A., 2001. Deep democracy: urban governmentality and the horizon of politics. *Environment and Urbanization*. 13, 23–43.

Benson, M., Jackson, E., 2013. Place-making and place maintenance: performativity, place and belonging among the middle classes. *Sociology*. 47, 793–809.

Bhat, M.A., Rather, T.A., 2013. Youth transitions in Kashmir: exploring the relationships between habitus, ambitions and impediments. *South Asia Research*. 33, 185–204.

Bourdieu, P., 1990. *The Logic of Practice*. Stanford University Press. Stanford, California.

Bourdieu, P., 2005. Habitus, in: Hillier, J., Rooksby, E. (Eds.), *Habitus: A Sense of Place*. Ashgate Publishing. Farnham, pp. 43–49.

Bourdieu, P., Passeron, J.C., 1990. *Reproduction in Education, Society and Culture*. SAGE. California.

Butler, T., Robson, G., 2003. Negotiating their way in: the middle classes, gentrification and the deployment of capital in a globalising metropolis. *Urban Studies*. 40, 1791–1809.

Centner, R., 2008. Places of privileged consumption practices: spatial capital, the dot-com habitus, and San Francisco's internet boom. *City and Community*. 7, 193–223.

Desai, A., 2002. *We Are the Poors: Community Struggles in Post-Apartheid South Africa*. Monthly Review Press. New York.

Dlamini, J., 2009. *Native Nostalgia*. Jacana Media. Auckland Park, South Africa.

Gale, F., 2005.The endurance of Aboriginal women in Australia, in: Hillier, J., Rooksby, E. (Eds.), *Habitus: A Sense of Place*. Ashgate Publishing. Farnham, pp. 356–370.

Gale, T., Parker, S., 2015. Calculating student aspiration: Bourdieu, spatiality and the politics of recognition. *Cambridge Journal of Education*. 45, 81–96.

Gevisser, M., 2008. From the ruins, in: Nuttall, S., Mbembe, A. (Eds.), *Johannesburg: The Elusive Metropolis*. Duke University Press. Durham, North Carolina, pp. 317–336.

Gevisser, M., 2014. *Lost and Found in Johannesburg*. Farrar Straus Giroux. New York.

Gotz, G., Todes, A., 2014. Johannesburg's urban space economy, in: Harrison, P., Gotz, G., Todes, A., Wray, C. (Eds.), *Changing Space, Changing City: Johannesburg After Apartheid*. Wits University Press. Johannesburg, pp. 117–136.

Hillier, J., Rooksby, E., 2005. Introduction to first edition, in: Hillier, J., Rooksby, E. (Eds.), *Habitus: A Sense of Place*. Ashgate Publishing. Farnham, pp. 19–42.

Holt, L., 2008. Embodied social capital and geographic perspectives: performing the habitus. *Progress in Human Geography*. 32, 227–246.

Huchzermeyer, M., 2011. *Tenement Cities: From 19th Century Berlin to 21st Century Nairobi*. Africa World Press. Trenton, New Jersey.

JHC, n.d. Tenant profile – inner-city of JHB: an overview 2006–2012. (Unpublished.) Johannesburg Housing Company.

Kihato, C., 2013. *Migrant Women of Johannesburg: Everyday Life in an In-Between City*. Palgrave Macmillan. New York.

Kihato, C.W., 2014. Lost dreams? Tales of the South African city twenty years after apartheid. *African Identities*. 12, 357–370.

Landau, L.B., 2006. Transplants and transients: idioms of belonging and dislocation in inner-city Johannesburg. *African Studies Review*. 49, 125–145.

Landau, L.B., 2018. Friendship fears and communities of convenience in Africa's urban estuaries: connection as measure of urban condition. *Urban Studies*. 55, 505–521.

Lefebvre, H., 1991. *The Production of Space*. Blackwell Publishers. Oxford.

Malcomess, B., Wilhelm-Solomon, M., 2016. Valleys of salt in the house of God: religious re-territorialisation and urban space, in: Wilhelm-Solomon, M., Nunez, L., Kankonde Bukasa, P., Malcomess, B. (Eds.), *Routes and Rites to the City: Mobility, Diversity and Religious Space in Johannesburg*. Palgrave Macmillan. London, pp. 31–60.

Mokonyama, M., Mubiwa, B., 2014. Transport in the shaping of space, in: Todes, A., Wray, C., Gotz, G., Harrison, P. (Eds.), *Changing Space, Changing City: Johannesburg After Apartheid*. Wits University Press. Johannesburg.

Oldfield, S., Greyling, S., 2015. Waiting for the state: a politics of housing in South Africa. *Environment and Planning A*. 47, 1100–1112.

Parnell, S., Pieterse, E., 2010. The 'right to the city': institutional imperatives of a developmental state. *International Journal of Urban and Regional Research*. 34, 146–162.

Prigge, W., 2008. Reading the urban revolution: space and representation, in: Goonewardena, K., Kipfer, S., Milgrom, R., Schmid, C. (Eds.), *Space, Difference, Everyday Life: Reading Henri Lefebvre*. Routledge. New York, pp. 46–61.

Rérat, P., Lees, L., 2011. Spatial capital, gentrification and mobility: evidence from Swiss core cities. *Transactions of the Institute of British Geographers*. 36, 126–142.

Sen, A., 2001. *Development as Freedom*. Oxford Paperbacks. Oxford.

Simone, A., 2008. The politics of the possible: making urban life in Phnom Penh. *Singapore Journal of Tropical Geography*. 29, 186–204.

Stanek, Ł., 2008. Space as concrete abstraction: Hegel, Marx, and modern urbanism in Henri Lefebvre, in: Goonewardena, K., Kipfer, S., Milgrom, R., Schmid, C. (Eds.), *Space, Difference, Everyday Life: Reading Henri Lefebvre*. Routledge. New York, pp. 62–79.

Stanek, Ł., 2011. *Henri Lefebvre on Space: Architecture, Urban Research, and the Production of Theory*. University of Minnesota Press. Minneapolis, Minnesota.

Thrift, N., 2005. But malice aforethought: cities and the natural history of hatred. *Transactions of the Institute of British Geographers*. 30, 133–150.

Winkler, T., 2008. Reimagining inner-city regeneration in Hillbrow, Johannesburg: identifying a role for faith-based community development, in: Harper, T.L., Gar-On Yeh, A., Costa, H. (Eds.), *Dialogues in Urban and Regional Planning*. Routledge. New York, pp. 133–149.

Zipin, L., Sellar, S., Brennan, M., Gale, T., 2015. Educating for futures in marginalized regions: a sociological framework for rethinking and researching aspirations. *Education Philosophy and Theory*. 47, 227–246.

6 The space that regeneration makes

Regulation, security and everyday life[1]

Introduction

The ambiguities of the regeneration process do not only make themselves felt in the ways people experience Johannesburg's unequal landscape and social order. Hierarchies, constrained forms of agency and limited capacities to alter urban environments are also inscribed into the ways in which people inhabit residential buildings. As this chapter will document, inner-city housing developments are tightly controlled spaces. Strict regimes of security and surveillance, whilst appreciated for the peace of mind and safety which they afford tenants, are used to discipline and control them, ensure that rental is collected and housing companies' authority is not challenged. These strict regimes frequently infringe on legal protections granted to tenants and demonstrate how commercial concerns take precedence over citizens' rights. However, the control exercised over tenants does not foreclose opportunities for socialisation and friendships. Tenants construct everyday lives for themselves which allow them to experience forms of solidarity and community, even in the face of harsh living conditions, economic necessity and intense forms of regulation. As they do so, they appropriate and inhabit inner-city spaces, and contribute to creating new meanings for and experiences of the city. Whilst some forms of appropriation take place despite the regulations imposed on tenants, others are facilitated by the regeneration process, and again demonstrate the multiplicity which urban regeneration is characterised by and giving rise to.

Regulation and control in inner-city buildings

A sense of security

All renovated housing developments are tightly controlled and highly secured. The entrances are fortified with metal turnstiles which can only be opened using electronic tags or fingerprint readers, as seen in Figures 6.1 and 6.2, and are watched by security guards 24-hours-a-day. Visitors seeking entrance to buildings have to sign in with security guards and the people they are coming to visit must

Figure 6.1 Entrances to social housing buildings. Photographs by the author.

Figure 6.2 Entrance to an affordable housing building. Photographs by the author.

come downstairs to collect them. It is therefore impossible to gain entry to buildings without tenants' and security guards' consent.

Whilst these systems appear draconian, residents tend to welcome them and point them out as the features of the buildings that they appreciate the most. In interviews, several tenants enthused about the security arrangements: For example, one tenant who lives in a social housing building in Hillbrow exclaims, 'It's very safe; the security access system, the lifts work, it's all fine. It's very clean. Where I was living before was not good, people got killed, mugged' (Tenant Seven, Lake Success). Tenants thrive on living in controlled environments and gain assurance from the security guards stationed in their buildings. As one tenant in a for-profit building in the CBD points out, 'I feel safe, nothing can happen. The security guards are nice people, they don't allow anyone in' (Tenant Two, Cavendish Court). Another tenant living in social housing illustrates the peace of mind tenants gain from security guards being present: 'The security is downstairs, we can sleep with the door open' (Tenant Fourteen, Gaelic Mansions). Thus, even though they are disciplinary figures who control the buildings and determine who is permitted entry, security guards are appreciated and the services they provide help tenants feel secure and at home. Although literature about the privatization and securitization of urban space mourns the loss of opportunities for chance encounters and spontaneity (Fyfe, 2004; Low and Smith, 2006; Mitchell, 1995), the narratives here point to the ways in which vulnerable people often prefer controlled spaces and that unpredictability, rather than being empowering or liberating, is often a source of anxiety.

It is, therefore, important to understand the designs and security features of these buildings as part of a South African vernacular lexicon regarding space and safety. Whilst the fortifications and access control certainly are 'draconian,' as even one housing company's Operations Manager admits, and call to mind defensive, militarised forms of urbanism (Caldeira, 2000; Graham and Marvin, 2001), they are also elements which contribute to people's sense of home and protection. Less contextually aware accounts of these buildings have stressed their militaristic nature and decried the way they resemble architectures of exclusion (Murray, 2011). Whilst this judgement does have some salience, it is also one that is based on an aesthetic imported from elsewhere and which does not take the vulnerabilities and lived habitus of inner-city residents into account. In defence of the security measures adopted, the Operations Manager argues, 'If you were staying in the inner-city you would understand it. And if you'd experienced it you'd understand it.' Because the inner-city is a volatile, violent and crime-ridden area, tenants welcome being able to retreat into places of security and comfort. The guarded entrances also allow parents to have peace of mind, as they know their children are supervised whilst they are away at work. As one tenant happily points out, 'Most of us spend the whole day at work and I know nobody is going to touch my daughter while I am away. The security is tops!' (Tenant Eleven, Gaelic Mansions).

However, to point out their localised significance and meaning is not to praise these forms of regulation and fortification; it is, rather, to call for contextualised understandings and appreciations of these types of interventions. Like appreciation for security and policing practices, dispositions towards safety are place- and context-dependent and are reproduced through people's lived realities. Hence in the inner-city, heavy security becomes the determinant of safety and gauge by which it is measured. As one tenant explains, 'Only when there is someone that is protecting the building downstairs, I'm very safe' (Tenant One, Rochester). A tenant echoes this desire for security when she declares, 'What I like about the place where I'm staying, I need to be safe, I need to be secured. That's what I like most about this place' (Tenant One, Rochester). It thus becomes clear that security, as much as it imposes orders and discourses on spaces, also reflects the prevailing forms of habitus and local contexts in which it occurs.

Enforcing rule by the market

But, bearing in mind duality and ambiguity as the hallmark features of the regeneration process, it remains vital to point out that the security features in residential buildings certainly are not benign. Housing companies also rely on security to regulate their tenants and create restrictive environments. The numbers of people living in apartments are tightly controlled and housing companies (both social and for-profit) do not allow guests to stay overnight without pre-arranged permission. Whilst this access control is generally welcomed by tenants, it can also restrict their abilities to assist friends or family members who need temporary places to stay, and is at odds with the fluid circumstances of many people living in the inner-city (see Mayson and Charlton, 2015). It is, therefore, a measure used to bring order and predictability into circumstances which frequently defy such ambitions.

The emphasis on access control is also part of a regime which is at pains to enforce rental collection and ensure that the commercial principles which define housing provision in the inner-city are not disrupted. Prospective tenants are screened very thoroughly. In order to secure accommodation, tenants have to provide proof of employment or proof that they have regular incomes. In the absence of letters from an employer, an affidavit signed by the police specifying the amount a prospective tenant earns per month must be provided. New tenants must also pay one month's rent up front as well as a deposit, which is usually equivalent to another months' rent; all companies insist that payments must be made through banks and do not deal in cash. In addition, tenants must have official South African identification documents or asylum seeker or residence permits. Although these measures are standard practice in almost all formal rental markets, they exclude large sections of the inner-city community, who survive on much smaller and sporadic incomes and frequently lack formal identification documents and access to bank accounts. They also create a situation where commercial arrangements and housing companies' authority go unchallenged and are consequently parts of efforts to order the inner-city.

In the recent past there have been intense struggles around housing and control of buildings in the inner-city. Landmark court rulings governing eviction processes and ensuring that people cannot be evicted if adequate alternative housing is not readily available have come into force (Tissington and Wilson, 2011; Wilson, 2010). There has also been an evolution in the framework governing rental arrangements and protecting tenants from unlawful eviction (SERI and CUBES, 2013). Prior to these developments, rental boycotts, active tenant committees, illegal evictions and building hijackings[2] were widespread and shaped the spatial politics of the inner-city (Beavon, 2004; Morris, 1999a; Murray, 2008). Building hijackings, in particular, continue to haunt the imaginations of housing providers and play prominent roles in shaping their management practices.

Housing companies have formulated their management strategies according to these 'institutional memories' and make sure that they have close control over what happens inside their buildings. Housing supervisors live in the buildings and interact constantly with tenants. They are generally chosen for their inter-personal skills and abilities to relate to but still exercise authority over the tenants and they keep close watch over the mood and levels of satisfaction inside buildings. Companies also use strict measures to ensure that rental collection is continuous and tenants have no option but to pay. Individual evictions, which have to be contested before a court or the Rental Housing Tribunal, are relatively difficult, costly and time-consuming to secure. They are, therefore, the measure of last resort. To avoid getting to a stage where evictions are necessary, housing companies institute a variety of punitive measures to make sure their rent is collected. Tenants who fail to pay their rent are first issued with a written warning, then have their lights switched off as further warning and are locked out of their apartments if they still fail to pay after receiving a final letter of demand. According to one interviewee, on average the company he is employed by switches off lights in approximately 100 apartments a month. The legality of these practices is dubious, as the Rental Housing Act stipulates that a landlord may not cut off water and electricity without a court order (SERI and CUBES, 2013). However, tenants interviewed did not complain about or query them, indicating that the housing companies enjoy high levels of authority inside their buildings and that citizens' rights remain difficult to realise and defend. It is a clear indication that dominance over space imbues actors with significant forms of power, and that the 'laws of the building,' as one building manager terms them, take precedence over legal frameworks.

Access control is also part of this regime and is used to enforce rental collection, discipline tenants and single out those who are behind on their rent. Again, with little regard for the law (which specifies that landlords may not lock tenants out of their apartments without first obtaining court orders (SERI and CUBES, 2013)) some companies employ strategies where, if a tenant is behind on their rent, their access to the building is deactivated. Once this happens, they are forced to go to the company's rental office and start making payments on the outstanding amount before they will be let in. Although tenants are not expected to settle all arrears immediately, the process makes sure that the company is monitoring the tenants and that payments continue. This not only helps avoid evictions and allows rental collection to carry on relatively seamlessly, it also makes each individual tenant subject to scrutiny and places companies' commercial considerations ahead of legal procedures and tenants' hard-won rights.

Access control mechanisms, then, are powerful reminders that spatial capital entails the ability to physically design and make places, and, in so doing, to assert particular interests and forms of order. In the South African context, the prevailing order cannot be separated from histories and contemporary realities of white supremacy and control over black bodies. Thus, although security features are welcomed by tenants and contribute to their feelings of safety, they can also be regarded as mechanisms which express the dominance of businesses, which are predominantly owned by white people, over spaces and the lives of (black) people residing in them. The memories and fears of building hijacking, whilst real for practical reasons, also echo fears about the unruliness of black tenants and the threat black lives posed to the white urban order (Erwin, 2015; Morris, 1999b; Popke and Ballard, 2004). One property developer talks about the access control mechanisms which have been put in place as part of an effort to create 'a culture of payment.' This 'culture' refers to an unquestioning acceptance of appropriate ways of behaving, respect for private property relations and an acknowledgement of tenants' place in the inner-city hierarchy.

Spaces of control and domination

Access control management and enforcing a strict regime of rental collection are techniques which help separate inner-city residential buildings from other 'unruly' spaces in South Africa, and assert their belonging to an order of regularity, financial discipline and submission to the rule of the market (and those who enjoy dominant positions within it). The interviewee quoted above contrasts the buildings his company manages with his imagination of Soweto and explains 'It's not like Soweto, where they'll start toyi-toying, burning tyres etc. [in response to efforts to enforce rental collection and payment for services].' Historically, housing in Soweto was built by the state, and thus withholding rents was a direct way to resist the illegitimate government which was laying claim to them (Crankshaw et al., 2000; Tomlinson, 1999). This refusal often endures in the contemporary period (Naidoo, 2007; Netswera and Phago, 2009) and the township's sprawling spatial form and large amounts of informal housing make it much harder to regulate and single out households who have not paid municipal bills or rent. In contrast, in the inner-city, because the laws of the market and private property prevail, as the aforementioned developer explains, 'There's a culture in the inner city that you pay your rent … You pay your rent, or you go somewhere else.' By submitting themselves to these conditions, tenants come to accept and be naturalised into an order of regulation and domination and submit to having their movements and everyday lives subject to constant discipline and surveillance. Access control is a physical tool used to achieve this and emphasises how controlling space is a way in which social and economic orders are enacted on a daily basis, and relations of domination become inscribed into everyday experiences. Triumphantly, the developer declares,

> If you take our biggest building, we've got 940 units in a single building, but you can *absolutely* control your access control; we've got biometric systems, security guards sitting downstairs, and the ability to deal with individual tenants is that much easier.

The South African vernacular order, in which black bodies continue to be acted on and moulded by the demands of white domination (Ally, 2011; Ndlovu-Gatsheni, 2013), therefore continues to shape the regeneration process in significant ways.

The management and regulation mechanisms put in place insist on the passivity of tenants and their submission to housing companies' rules and laws. In addition to being naturalised through the habitus and lack of spatial capital which tenants lay claim to, their acquiescence is also actively enforced by the stipulation that tenants may not form committees inside residential buildings. Aside from one, all the companies in whose buildings interviews were conducted do not allow tenants to form committees and insist on dealing with them as individuals. This is a violation of the rights afforded to tenants as spelled out by the Gauteng Unfair Practices Act (SERI and CUBES, 2013). Whilst companies are open to suggestions about ways to improve the buildings, they are adamant that committees will not be recognised, particularly if they venture into challenging economic arrangements. As one private developer explains,

> We don't encourage tenant committees because they tend to flare up when there are service interruptions or difficulties in the buildings and it becomes a platform for a whole variety of a shopping list of issues and it often becomes very political and polarised. Obviously everyone wants to have free housing and accommodation for nothing, but commercially the building needs to be viable financially and to run at a profit because it is a private sector enterprise.

Although the majority of tenants interviewed were unconcerned about the lack of avenues for voicing complaints collectively, and may prefer anonymity, not having to engage with their neighbours and strict management which, although infringing on some of their liberties, keeps the buildings peaceful and running smoothly, some tenants did complain about being prevented from organising inside the buildings. They note that these practices contribute to feelings of estrangement and passivity. For example, one tenant complains that 'Everybody minds his business in this building. We are not united as tenants, we don't even have tenant meetings' (Tenant Four, Greatermans). Similarly, another bemoans the way the management company running the building he resides in refuse to deal with tenants on a collective level: 'Trafalgar, they don't want us as tenants to meet, converge, convene, compare notes and go to them as a team; they want us to come individually. And that's something I don't like' (Tenant One, Ridge Plaza, Berea). Whilst tenants are not prevented from meeting in alternative spaces, the knowledge that their committees will not be recognised discourages them from doing so. Housing companies, then, utilise their spatial capital and dominance to define both the issues which tenants are allowed to query and voice concerns over, as well as the ways in which they socialise and interact with one another. They therefore use this power to express and naturalise their dominance within the inner-city milieu.

Inhabiting and appropriating space, transforming the inner-city

Solidarity, friendship and everyday agency

However, although forms of regulation, spatial ordering and the uneven distribution of power and capital are all features of life in the inner-city for tenants, and express the ongoing marginalisation and inequalities which continue to characterise post-apartheid transformation, experiences of inhabitation, change, agency and everyday life remain complex and have multiple expressions and consequences. Abilities to make or take place, and thus exercise spatial capital, are also expressed through the everyday and the ways in which people, in making themselves at home in particular spaces, infuse those spaces with meaning and change what they represent (Purcell, 2002). The movement of black people into the inner-city, a formerly white, segregated space, is in itself infused with political significance and a form of transformative power. One tenant captures this powerfully and situates black people's occupation of the inner-city with the freedom brought by the end of apartheid. He speaks emotionally and recalls,

> When there was this new political dispensation in 1994, the people – I'm talking about black people – started moving from the townships. Those who were wealthy from township standards started moving into places like Berea; Berea, Hillbrow is first place where black people came when they were coming from townships. So it was like a *wow* thing, now suddenly we had this freedom, now we could live in these buildings which previously black people were denied! So it was unbelievable!
>
> (Tenant One, Ridge Plaza)

The newly realised freedom of living in the inner-city was, unfortunately, curtailed in many ways by the anomie which the area descended into in the immediate post-apartheid period. However, through the regeneration of both living spaces and the public spaces surrounding buildings, new meanings and expressions of this freedom are emerging on an everyday basis. Environments conducive to family living not only allow working-class black households to secure their social reproduction, they also change what these spaces represent. Buildings with names like Gaelic Mansions, Rochester, Cavendish Court and Dorchester, which were originally built and named to signify relations of

coloniality and exclusion, have now come to serve as fulcrums of black communal life. They have thus undergone a transformation in what they stand for and
what they make possible, calling attention to the spatial capital which people
possess and which allows them to inscribe meanings into places, by virtue of
their presence in them.

Processes of making home are also ways of exercising spatial capital and
abilities to make place. This is evident in the way in which, through forming
social networks, friendships and support systems, tenants have been able to
transform (on some level at least) the ways they experience the inner-city.
Whilst for many it does still represent (for good or bad) a place of anonymity
and distance from others, for some it is a place where friendships and mutual
support make life easier. Although the majority of tenants did not share these
experiences, some tenants did speak favourably about the friendships they have
cultivated in the buildings in which they live. For example, although most
responded in the negative when asked if tenants in the building socialise, two
social housing tenants reply:

> Quite a lot – there are sports teams in the buildings, courses to bring
> teenagers together, in-house sports, pre-school and a crèche. I support the
> netball teams and do talks and career advice with teenagers.
>
> (Tenant Six, Lake Success)

> We all know each other, we can play soccer together, children can do
> anything. People participate; we elders have two hours playing on Sundays.
>
> (Tenant Three, Lake Success)

Whilst these responses were not the norm, and, as mentioned earlier, some
tenants were not even aware of the social programs and childcare facilities
available, they do point to emerging forms of everyday cohesion and communal
solidarities. These solidarities, improved conditions created by urban regeneration efforts and the new-found mobility and choice afforded to people are
allowing them to establish more settled, stable and comfortable ways of being
in the inner-city. Indicating the break with apartheid-enforced geographies, one
tenant reflects, 'I've tried going and living in Soweto and came back. I only feel
comfortable here in this area' (Tenant Two, Greatermans). Another emphasises
that the ways in which people are able to forge new personal relationships in the
spaces in which they are living assist them in feeling at home and dealing with
the harsh environment they find themselves in: 'Ja, it does [feel like home], even
though my family is not here, but with the people that we live with, I have
friends around, I think it feels like home' (Tenant Nine, Rochester).

Importantly, these friendships alter the affective atmosphere of the area for
some, and change the way the space is experienced. For instance, one building
manager relates an event which took place in the building he runs shortly after he
moved in. He explains how two tenants were taken ill and how their neighbours
rallied around to care of them, cooking food, looking after their children and

even raising money for them to travel back home to the Northern Cape. This experience not only surprised him, but changed his opinion of the building and the people living there. As he reflects,

> I got that sense that there is a community here, because I had only been here for eight months or something, but that sense of community, that sense of wanting to help each other, I saw it that time and I said 'People, they are living in a community here.'

These experiences and actions thus become expressions, albeit mundane, of tenants' spatial capital and abilities to make spaces which can accommodate them. As Butler and Robson's (2001, 2003) work in London demonstrates, spatial capital is augmented by social capital. Networks, shared dispositions and values and opportunities to socialise all play roles in shaping the ways in which neighbourhoods change. Thus, knowing each other, having opportunities to socialise and forming support networks (i.e. improved forms of social capital) play important roles in assisting tenants navigate a harsh environment and help them make homes in it. Therefore, as infrastructures are put in place, social programs are enacted and spaces and opportunities to socialise increase, so too do tenants' abilities to make themselves at home in the area, and thus make the area itself. Through their forms of social support, care and maintenance, tenants help make urban life possible for one another and bring 'the good city' (cf. Hall and Smith, 2015, p. 5) into being.

Relationships between tenants also become part of the reflexivity and contingency which define management in the regeneration process. For example, although social events are not permitted in buildings and the use of communal spaces is tightly regulated, one building manager recalls how tenants were able to persuade him to deviate from the norm and allow them to hold a Christmas party in the building. He recalls, 'We don't allow such things in awkward times, but there's some times when you must leave tenants to enjoy themselves,' thus demonstrating how the contingencies and intricacies of daily life necessitate flexibility and adaptability. Through these measures, tenants are able to alter the strict arrangements in the buildings and find ways to have their needs recognised and accommodated. Reflexivity and relaxing the rules does not only occur during festive times, but in everyday interactions too. I have argued previously that adaptability and contingency are essential elements of the regeneration process, and that these dispositions are inculcated through people's experiences in the space and the way they adapt to it. It is also evident that tenants play a role in making contingency necessary and shaping the practices of housing providers. Although the arrangements regarding rental collection are stringent and unforgiving on paper, in practice the day-to-day relations and politics of sharing space make them somewhat less harsh. All of the building managers/housing supervisors who were interviewed, although acknowledging rental collection as their primary task, were also loath to actually enforce rules regarding lock-outs, and devised ways to avoid having to do so. As one manager (who was the strictest out of all the building managers I encountered) relates 'I make sure that I don't lock at night 'cause they will be stranded. I make sure that I wake up early in the morning and I tell them "They're coming to lock," then they will make sure that during the day they will pay.'

I was also able to witness the personal and reflexive side of building management in practice, when, during the course of one interview, a tenant came to the office in which the interview was being carried out and engaged the building manager in conversation. It was the 22nd of the month, and the tenant had finally been locked out of her apartment for not paying her rent. Not being able to avoid locking the tenant out, as this would have been in dereliction of her duties, the building manager arranged for the tenant and her daughter to stay in her apartment. She thus found a way to express personal sympathy with the tenant and alleviate some of her suffering, whilst still conforming to the demands of her job. It was an instance in which the personal and reflexive side of the regeneration project, as well as some of its harsher aspects, was brought home, and demonstrated the ways in which people make do and exercise agency, even within the confines of a strict, regimented process, and how these actions contribute to making the lived reality of the inner-city. These examples emphasise the importance of the everyday as a site in which new relationships, forms of solidarity, care and affection take place, and demonstrate how these are not isolated incidents, but form part of broader narratives and experiences of a changing, confusing urban space. As Back eloquently points out,

> Tales of social damage, hopelessness and injustice always make for a good sociological story. But the cost is we too often look past or don't listen to moments of the repair and hope in which a livable life is made possible. This is why an attention to everyday life matters.
>
> (Back, 2015, p. 13)

Building bonds and creating new social spaces

Since the end of apartheid, Johannesburg's inner-city has not only become a site of racial transition, it has also become an aspirational place for migrants from across the continent. One interviewee, who is originally from Zimbabwe speaks with a sense of awe and excitement when he relates,

> The whole of Africa when you talk about Johannesburg – not South Africa, *Johannesburg* – it's like you are talking about *the mecca* because this is where everything is happening, this is where they think there are opportunities, here in Johannesburg. So you'll hear them talking about Cape Town or Durban but everyone is talking about Johannesburg the most!
>
> (Tenant One, Ridge Plaza)

Vernacular approaches to regeneration

A building in the heart of Hillbrow exemplifies the vernacular, localised form regeneration has taken. The Old Synagogue (Figure 7.1) is located on Wolmarans Street. Built in 1914, it was the third synagogue constructed in Johannesburg and was the centrepiece of Jewish life in the city.[1] However, as white residents abandoned the inner-city, the synagogue fell into disuse. It was closed in 1994 and a replica was built in Melrose, an upmarket suburb, soon after. The original structure endures today, and has come to capture many elements which make up the new social order of the inner-city. Nuttall (2004, p. 744) illustrates how, for the novelist Pashwane Mpe, Hillbrow is 'figured as the partial and now patchy inventory of the old apartheid city and as the revised inventory of a largely black, highly tensile, intra-African multiculture.' The Old Synagogue shows this clearly. The original facade of the synagogue has been preserved, but in altered form. The iconic dome and Star of David still stand out clearly, but the outer walls of the building have been renovated and now accommodate various shops, which cater to the needs of the inner-city's new population. Shops selling cheap groceries, cell phones and related paraphernalia, a tavern and even a (non-kosher) chicken restaurant now thrive in the shadow of this once-sacred space. These changes show how inhabitants of the inner-city refashion the infrastructure they have inherited from the old order and adapt it to suit their needs.

Figure 7.1 The Old Synagogue. Photograph by the author.

Whilst the outer facade of the building serves material needs, the inside continues to nourish inner-city residents' spiritual lives. It is now the home of the Revelation Church of God, an offshoot of the Universal Church of the Kingdom of God, a major transnational church which was founded in Brazil but has spread rapidly around the world, particularly on the African continent (Malcomess and Wilhelm-Solomon, 2016). Twice a week, on Tuesdays and Sundays, huge crowds gather inside the building and eventually spill over into the streets surrounding it, as rows upon rows of faithful worshippers gather for ecstatic church services. The original structure of the synagogue has been kept intact and inside plaques still honour members who made contributions to its construction and maintenance over the years. However, as the old users have moved on and established new communities elsewhere in the city, new users have taken their place and adapted the remnants they left behind to suit their own needs and circumstances. Worshipers sing elated gospel songs that rock their bodies, as well as the old but still solid structure, and those in need come forward to be healed by the congregation's charismatic leader. Whilst this is taking place, Bad Boyz security guards – whose head office is located on the ground floor of the high-rise apartment building opposite the synagogue/church (seen in the background of Figure 7.1) stand by, making sure the crowd is kept under control and no untoward incidents take place. In this space, then, the different worlds and social orders which constitute the inner-city come together. Privatised, securitised urbanism mixes with syncretic, globalised Pentecostal Christianity, in the shadow of a building built by a white, European immigrant population in the centre of an African country, in the heart of what was once, and for many still is, an unwelcoming, hostile part of the city (Parker, 2017). It is a visible and palpable 'piecing together of aspects of city life – people, things, spaces – that are not conventionally thought to be associable' (Simone, 2010, p. 151), but which is usurping the dominant representations, experiences and meanings which Hillbrow has been burdened with. It demonstrates the different dynamics which characterise the regeneration process coming together simultaneously, and the novel, surprising vernacular experiences they give rise to.

Whilst some of the issues I have dealt with in this book are more prosaic than the scene depictedabove, they too represent the vernacular, at times surprising, and generally novel combination of processes, materials, agendas and practices which define the regeneration process. These experiences, although encapsulating a particular set of dynamics, historical trajectories and contemporary realities, need not be considered as solely relevant for the physical setting in which they are taking place. The various chapters have all shed light on issues, processes and forms of agency which are relevant to and relatable for a number of other settings and societies. All the processes documented do not take place outside of international trends and pressures, but engage directly with them, although in localised ways. For example, in trying to define urban regeneration and explaining how it should be done, the Johannesburg Housing Company's (JHC) Chief Executive Officer references the current international policy vogue for mixed-income communities. However, as she explains, it takes on a specific form in the South African context. Thus, she states,

> If you do urban regeneration you must be very careful that you don't push poor people out, because if you gentrify a city too much and push out poor people, that's not urban regeneration, in my view. So, it doesn't help for us to strive towards bringing in high-end people in the inner-city and Manhattan-type developments and they don't want to live alongside poor people. So, if you do urban regeneration within a precinct there needs to be a bit of everything, and that's what we try and do in our buildings so that there's not poor people living in one block and high-end people in another block.

High-end housing has not been successfully developed in the inner-city. Rather, the 'mix' that she is referring to describes a situation in which housing catering to households in the low-income, social and affordable housing brackets (i.e. households earning less than R3500 per month, between R3500 and R7000 per month and R7000 to R14 000 per month respectively) exists together in one building. Mixed-income in this case is thus not a term used to sugar-coat the harmful effects of gentrification and displacement of the poor, as has been the experience in other contexts (Butler and Lees, 2006; Chaskin and Joseph, 2013; Colomb, 2007; Lees et al., 2016; Smith, 2002). Processes of urban change do indeed follow global trends, but these are inflected with localised concerns and attempts to avoid recreating failed or segregated forms of urbanism seen elsewhere. Johannesburg's post-colonial urbanism is not only mimicry (Mbembe, 2008), but actively learns from and improves on experiences from elsewhere, in light of the prevailing local conditions.

Contingent, adaptive practices which blend international trends with localised imperatives are also discernible in the ways capital accumulation proceeds and developers operate. For instance, when articulating his long-term ambition or vision for the inner-city, a for-profit developer draws firmly on a neoliberal habitus and shapes his response through his position as an entrepreneur. At the same time, however, he hybridises this habitus with a social commitment born out of and in response to the socio-spatial conditions which prevail in the inner-city. Hence, he declares,

> We want to be if not *the*, certainly perceived as one of the premier rental housing businesses who are offering good, solid accommodation, who look after their tenants properly and we're doing a fair deal and we're running a fair business. That's what I want to achieve, and it's very unsexy and very boring but you get it right and the blooms will come.

Again, this illustrates the ways regeneration is focused on achieving incremental changes and making the area liveable, rather than pursuing grand visions of urban upgrading. However, grander visions are not absent, but again they are formed in response to the realities of the area and the needs of the people inhabiting it. Thus, he elaborates on his long-term visions by stating:

> We get that right [the provision of basic accommodation and the stabilisation of the area] and in two years' time we've got a 2000m² retail outlet, high-end, proper, not a spaza shop [a trading kiosk], a proper outlet ... You have that, you have some other line stores, maybe take these offices and convert them, do something like put an AIDS clinic in here, a business incubation centre. I'll provide the space, I want Liberty Life [a major insurance company] and TUHF to sponsor the computers ... let's get some community stuff going ... So I'm again trying to create mixed-use, hopefully this is a piazza now, suddenly we can get some traction.

Here he shows that visions of public space derived from elsewhere are adapted to local needs and conditions, and are hence vernacularised. Whilst the neoliberal habitus comes to the fore, as business and entrepreneurialism are heralded as the solutions to poverty, he also shows social awareness and cognisance of the potential for regeneration to respond to the prevailing social problems. Commercial needs thus sit alongside developmental needs in his and other developers' visions, demonstrating the ways in which the process is not another iteration of a global trend or phenomenon, such as gentrification or neoliberal urbanism, but is a real attempt to blend contrasting approaches and create a city which responds to the needs of its inhabitants.

In some cases, these needs are dire and immediate, and regeneration cannot afford to have grand ambitions, as it must tackle critical and systemic neglect and a decayed urban environment. A housing supervisor draws on the prevailing physical conditions in the inner-city and explains regeneration as an effort to improve the area. In his view urban regeneration means

> maintaining the infrastructure here in the city, taking care of those abandoned buildings, renovating them and making them habitable – a place that humans can go and habitate [sic] them, that's what urban regeneration means to me. Making the whole city habitable.

Clearly, the particular spatiality of the inner-city has shaped the ways in which actors respond and is thus embedded in their habitus. At the same time, responding to these harsh conditions entails more than physical upgrading and has results beyond the built environment. Just as the spatial landscape inculcates ways of thinking, seeing and acting, improvements made to it also extend further and take on broader socio-political and symbolic significance. Thus, another housing supervisor demonstrates the social commitment which is at the heart of the regeneration process and how this tempers commercial concerns. Significantly, he points out how social commitments include helping people establish senses of belonging and community:

By regenerating the inner-city, we are making it better and taking on new buildings that are old, making them new and trying to make people being comfortable and welcoming them into buildings. Then you will make them feel better. And your rental settings, also it counts a lot ... when you do your rental settings, they must be affordable. You don't just become a skyrocket [sic], then your buildings will stay empty because people can't afford that.

The ways in which people are establishing community and finding belonging in the area are varied. In some cases, they have failed to do so, or chosen not to, and remain detached and resigned to living in the area as it is the only available, but far-from-ideal, option. However, evidence has also been presented of new forms of solidarity and social cohesion emerging, including through policing practices, volunteering, participating in events organised by housing companies and their associates and everyday social life in residential buildings. All these activities are creating new forms of associational life in an area that was initially segregated and off-limits to most black people and was then rendered largely inhospitable through widespread governmental neglect and capital flight (Gotz and Simone, 2003; Winkler, 2013). Through the collective efforts of a variety of actors it is now an increasingly welcoming environment and one which symbolises significant transformation and potentiality in the post-apartheid social landscape. The housing supervisor quoted above captures the energy, dynamism and transformation which regeneration promises. He declares,

Urban regeneration means to me bringing new life in this city, that's my understanding. Because when you regenerate a thing you are making it new and bringing new ideas, new types of buildings and new environment. But all in all, it's bringing new life into town.

The new life which is springing up in the inner-city is testament to this and demonstrates the changes underway in the post-apartheid period. Whilst there are several shortcomings and areas of concern, the regeneration project has been central to the emergence of new forms of urban life. It thus straddles and signifies multiple forms of social order, including the embedding of neoliberal practices and forms of governance as well as the fostering of a developmental, transformative order. It is thus a process which needs to be understood as a vernacular response to both localised, specific conditions and broader global trends and currents.

Vernacular as universal condition

Comaroff and Comaroff (2012, p. 9) suggest that 'African modernity is a *vernacular* ... wrought in an ongoing geopolitically situated engagement with the unfolding history of the present' [italics in the original]. Given that the inner-city is one of the key spaces through which this modernity is unfolding (Mbembe and Nuttall, 2008), it follows that the inner-city is itself a vernacular space, constituted by local practices and forms of habitus, which themselves are situated in and emerge out of wider geopolitical contexts. The inner-city is therefore a particular space, but not one which exists in isolation. I wish to argue that all spaces have their own vernaculars, which are formed out of the comingling of distinct and shared histories, cultures, structuring forces and practices. It is the combination of that which is specific to a particular context and the broader, shared processes and dynamics that I want to draw attention to, and urge scholars working elsewhere to address too.

Theorising urban change from a vernacular perspective means employing a variety of vocabularies and conceptual registers and allowing for varied, unexpected findings and outcomes to emerge. It also means being attuned to the uncertainties, multiplicities and constant forms of becoming which define urban spaces. Post-colonial societies are good, but certainly not the only settings in which this mode of thought can be experimented with. By nature, they are contested, fragmented, diverse and dynamic (Chatterjee, 2013; Comaroff and Comaroff, 2012; Robinson, 2003; Roy, 2009; Watson, 2014). This is not to dismiss the multiplicity of Western/Northern societies, which certainly are dynamic and home to varied experiences and ways of being urban (Dwyer et al., 2013; Hall, 2015a; Peck, 2015), but to recognise that forces and practices of experimentation, innovation, uncertainty and resistance are frequently closer to the surface in societies emerging from the traumas of, and in many ways still subject to, oppressive regimes and externally imposed forms of government and domination.

It is also important to bear in mind that distinctions between Developed and Developing societies are false binaries and overlook the ways in which, rather than being separate worlds, these societies are products of a shared, mutually constitutive history and remain joined together in the present (Robinson, 2006). The colonised world has long served as a site of experimentation for various policies and styles of governance emanating from and now being adopted in the West (Comaroff and Comaroff, 2012). For instance, the rounds of austerity being forced on some European Union states and voluntarily adopted in the United Kingdom in the wake of the 2008 financial crisis bear strong resemblances to the Structural Adjustment Programmes which were imposed on many countries in Africa in the 1970s (ibid.). Pinochet's Chile was also one of the first countries to be subjected to neoliberal shock therapy (Peck, 2015), illustrating that the South, rather than being backwards and always in a state of playing catch-up, is actually at the forefront of economic, political and social change. In terms of resistance, some draw lines of continuity between the public demonstrations and occupations which came to be known as the 'Arab Spring,' the running battles fought over public space in Istanbul, and the Occupy movements which arose in London and New York (Harvey, 2012; Merrifield, 2013). Whilst the similarities between the different movements are debatable, the Occupy movements certainly drew inspiration and lessons from the uprisings in other parts of the world, showing that the South can be a site of learning for the West and provides glimpses into emerging forms of social relations and political organisation across the world. Urban policies too are circulating rapidly, with cities in the South often being sites of learning for each other, as well as for the West/North (McFarlane, 2006; Wood, 2014). Southern or post-colonial societies thus need to be read not as Western societies' 'Other' or poor relative, but as offering insights into processes of change, hybridisation and social upheaval which are being experienced on a global scale.

Thus, a theoretical lens developed in and from the perspective of these societies does not only respond to local particularities or imperatives, but potentially speaks to a globalised world where Western and post-colonial societies are mutually intertwined in circuits of capital and crisis, policy formulation and transmission, environmental stress, processes of urban change/gentrification and flows and experiences of migration. The particularism of post-colonial societies stands for the particularism of all societies and contexts, and the insights, theorisations and epistemological perspectives gleaned from these settings are potentially applicable everywhere. In her 'ordinary cities' approach Robinson (2011, 2006) does not strive to reassert post-colonial particularity, but aims to develop a theoretical lens and register which can speak to and about intertwined and shared urban experiences. Her starting points may be Brazil and South Africa, but her objective is global – the world of cities everywhere. Theories of neoliberalism, gentrification, revanchist renewal, Global and World Cities and planetary urbanisation all have pretentions to global status – why should theory developed in post-colonial contexts be different?

On the one hand it should not be – post-colonial perspectives are not only able to articulate and appreciate the conditions prevailing in their contexts but are also able to speak to the conditions which are rapidly becoming global and points of crisis everywhere. On the other hand, post-colonial theory should be different as it is born out of experiences of domination and totalising narratives which it should be loath to replicate (Chakrabarty, 2009; Spivak, 1999). Thus, whilst post-colonial theories may speak to global experiences, or experiences which are becoming global, they should do so by highlighting the inescapable messiness and volatility of social and urban life everywhere. Starting from this basis, it becomes vitally important to be attuned to diversity, emergence and the constant states of becoming which define urban societies, as I have attempted to do. Modernity and urbanity in Africa are experimental, dynamic and defined by specific conditions emanating from colonial histories, but are also worldly and have always been formed by and contributed to the formation of Western capitalism, imperialism, modernity and urbanity too (Mbembé and Nuttall, 2004; Myers, 2011; Pieterse, 2011). One cannot think African urbanity without the West and the colonial experience, but the West's modernity relied heavily on and was shaped by relations of coloniality, and continues to do so today (Grosfoguel, 2007; Gutiérrez-Rodríguez, 2010; Maldonado-Torres, 2007). Can London's present forms of urbanism be grasped accurately without factoring in the influence migrants from across the post-colonial and post-socialist worlds are playing, not only in its property and labour markets, but also in its social formations, relations, and everyday life (Hall, 2015b)? Thus, 'multiplexity' (Simone, 2004, p. 214) is not only experienced in the South, but being orientated towards it and to thinking with variance and difference, and thus against monolithic or too-easily travelling concepts, can be learnt from the experiences and epistemic viewpoints emerging out of post-colonial contexts.

Within the vernacular framework I have adopted and argued for in this book, the salience of terms such as neoliberalism, gentrification and revanchist renewal is not dismissed. These are real driving forces shaping many urban settings, including inner-city Johannesburg. However, these terms do not capture all, or even most, of the forces, points of conflict, experiments and imperatives which are driving moments of urban change. A vernacular approach to understanding these moments thus views these critical concepts and the issues they highlight as important elements and outcomes, but also sees them as subject to being influenced by and changing in the face of local conditions and agency. They exist alongside other important political and social dynamics and attention to the combination of these and the novel insights which arise drives a vernacular approach.

In closing, I want to address some specific areas of concern which my research has shed light on, and which I hope can be helpful for developing new understandings across multiple sites and settings.

Governance and politics as dynamic, complex processes shaped by multiple agendas

Firstly, the research and findings I have outlined in this book speak to the dynamism, uncertainty and multiplicity of governance agendas and processes. As the arguments in chapters 1, 2 and 3 highlight, urban governance and policy formation in post-apartheid South Africa have been ambitious as well as conflicted undertakings. It is not possible to distil these efforts into a single agenda or outcome, as they comprise varied, frequently contradictory practices, goals, ideologies and ideals (Parnell and Robinson, 2012; Robinson, 2015). The policy landscape framing the regeneration process combines neoliberal, market-driven agendas and interests with genuine attempts to provide housing which meets the needs of low-income communities and contributes to social redress. It is also a (largely unsuccessful) attempt to balance the needs of one specific locality in the city with the demands of a wider, vastly unequal city-region. To make sense of all these competing, contradictory currents and agendas, we need to utilise a conceptual approach which understands governance and policy formation, and the actors behind these processes, as inherently complex, creative and heterogenous.

Narratives about global capitalist or gentrification agendas are powerful, and often provide valuable insights into the workings of the real estate-finance-state nexus (for example see Aalbers, 2012; Halbert and Attuyer, 2016; Lopez-Morales, 2011). But focusing mobilisations and forms of resistance outside of the state and overlooking actors within states who are pursuing alternative, socially progressive agendas not only oversimplifies a complex social terrain, but removes a powerful ally from efforts to ameliorate the harmful effects of property-and-finance-led urban redevelopment and build more just cities. Based on the lessons from Johannesburg, I therefore want to urge researchers to pay close attention to difference, resistance, agency and progressive practices or aspirations occurring *within* states and governance processes, and to re-centre the state as a possible site of social change, which needs to be reclaimed from neoliberal, opportunistic elites.

Property developers as socially embedded and reflexive actors

Following from the above point, I also want to highlight the need for research to be aware of variance amongst people who are driving investment and redevelopment processes. My research with property developers and housing providers in Johannesburg, presented mostly in Chapter 3, underlines how this group of actors, whilst being economically and spatially dominant, are also influenced by the different currents, values and forms of distinction which circulate in society. Cravings to reap profits from investment and housing provision certainly are central to their dispositions and practices but so are desires to make the inner-city a better, more welcoming environment, to contribute to poverty alleviation and to participate in the city's transition away from its racially exclusionary history. The dispositions which shape these powerful actors are outcomes of competing and contradictory policy logics and agendas and arise out of their interactions with the spatial realities of the inner-city. It is important, then, for research in other settings to engage with developers as socially and spatially embedded, and to understand the ways in which dispositions, forms of habitus and spatial praxis vary across different settings, and to understand and reflect on some of the causal factors behind variance.

At the same time, as property development becomes increasingly global and mobile, and developers move across contexts (for examples see Ballard et al., 2017; Zheng, 2013), research which explores the ways in which forms of habitus and spatial capital travel and adapt (or fail to do so) as actors move from one site to the next can be helpful for not only re-injecting agency and humanism into the workings and flows of international capital, but also for examining the relationship between various forms of governmental regulation and policy oversight, societal pressure and the ways in which urban redevelopment proceeds or can be altered. In this book I have sought to present a sociology of the field of urban regeneration in Johannesburg, and would urge others to examine urban regeneration/redevelopment/gentrification processes as defined fields in other settings too, and to unpack the animating logics, values, aspirations, habitus and forms of spatial capital which shape practices and potential outcomes.

Security as an everyday concern

Discussing policing, regulation and the control of public spaces by private security forces, I have also attempted to bring the ambiguity of these practices to the fore. Again, whilst in no way commending or justifying the practices and tactics employed in Hillbrow, I do want to reiterate the multiple meanings and outcomes which processes of securitisation have, and to underscore how, particularly in fraught, volatile, violent urban contexts, the search to find safety and security structures people's decisions and aspirations and can never be taken for granted. Ensuring security, even through heavy-handed and privatised means, is a powerful component of forming community and attachments to places, and therefore needs to be factored into any urban planning or development process. I would therefore call on researchers to adopt nuanced, ambiguous perspectives on urban security and policing, and, whilst remaining critical of regimes and forms of violence and exclusion, to bear in mind the multiple, contested meanings which these have.

In urban studies, the prevailing literature on policing, surveillance tends to present a picture of a relentless march towards homogenization and a loss of public space (Flusty, 2002; Fyfe, 2004; Mitchell, 1995; Németh, 2006). In postcolonial contexts, proximity to violence, disorder and even death are often part of everyday life, and publicness takes on varied, sometimes dangerous meanings (Alves, 2014; Mbembé, 2003). We therefore cannot operate with universalist, uncritical notions of public space, nor of what it means to create security in these spaces (Houssay-Holzschuch, 2016). Again, this stands as a call for difference, fluidity and vernacular perspectives to be taken seriously in thinking about processes, forms of belonging and places. At the same time, the harmful, necessarily exclusionary consequences of community formation, boundary drawing and policing cannot be overlooked, nor can we ignore the way in which privatised approaches have meant that crime has been displaced, rather than effectively dealt with. Experiences from Hillbrow clearly illustrate how development processes and security provision are incomplete as long as they fail to address the societal drivers of crime and violence, particularly poverty, lack of access to education and employment opportunities, gender-based violence and collective forms of trauma. The challenge therefore remains to find multi-faceted approaches to dealing with competing claims to urban space, diverse meanings of belonging and safety and contested notions of community and inclusion.

Inhabitation and everyday agency

Ambiguity and variance are also part of everyday life and the ways in which people experience and make homes for themselves in urban spaces. My research has revealed the ways in which residents in inner-city housing developments gain some benefits and new experiences of spatial capital, upward mobility and urban inclusion, whilst simultaneously having certain rights restricted and developing detached, resigned relations with the spaces in which they are living and the people they are sharing them with. At the same time, new forms of solidarity, friendship and associational life are also emerging on an everyday basis which provide the care and support which people require. Experiences are therefore varied and do not cohere into singular accounts or narratives. These findings draw attention back to everyday life as a site of creativity and adaptation, and demonstrate that agency is present in quiet, mundane forms as well as the choices people make, even when choosing from limited options. Thus, whilst moments of resistance remain vital, research on urban belonging, forms of habitation and claiming space needs to pay attention to the subtle, everyday forms of agency which help people endure in difficult and changing circumstances too. It is also important to understand some of the factors which shape everyday agency and often curtail more overt forms of resistance or politicised forms of appropriation from arising. Working through concepts such as 'spatial habitus' and 'spatial capital' can be a valuable way for doing so, as these concepts draw attention to the intertwined and mutually reinforcing relationships between social hierarchy, experiences in space and abilities to shape or take place.

As the political theorist Chantal Mouffe (2005) asserts, we need to pay attention to the political as well as spatial conditions which foster (or conversely hinder) the development of democratic, participative forms of habitus and citizenship. Planning processes and efforts at community mobilisation thus need to be cognisant of the spatial and structural forces which frequently hinder participation and political involvement in urban life, including long working hours, misrecognition of abilities to effect change, resignation, temporary forms of residence, desires to be elsewhere, and the leisure of sitting back and letting other people worry about things. At the same time, research also underscores the strategic value of disinterest, anonymity and being disengaged from the spaces and social relations in which one is living (Kihato, 2013; Landau, 2018; Landau and Freemantle, 2010; Simone, 2001). This presents a challenge for effective community consultation, participation and mobilisation. I do not claim to have a solution to these conundrums, but hope that providing some tools for re-focusing attention on the politics of everyday agency and mundane forms of inhabitation can contribute to thinking through some of these issues.

To end, writing this book, whilst based on detailed empirical data and evidence, has been an exercise in writing the Johannesburg of my imagination and aspirations into being. I have tried to make the Johannesburg of inventiveness, sociability, change and meaningful attempts to deal with stubborn problems and forms of exclusion real, and make it come alive amidst all the other versions of the city which exist – Johannesburgs of selfishness, exclusion, violence, failed governance, too-successful forms of neoliberal governance, segregation, white racism and dystopia. Writing about cities is always a form of creativity, as it conjures up images, visions and versions of reality. It is also always a form of violence, as it imposes one meaning onto a space and reality in which multiple experiences, narratives and realities exist. I therefore do not claim to present a definitive account of Johannesburg. But it is my hope that out of this necessarily selective process, a more hopeful and nuanced vision of the city can come to take prominence, and that it can resonate with and be a source of learning for people living and working in other contexts too.

Note

1 www.greatpark.co.za/pages/Great%20Park%20Synagogue%20History.

References

Aalbers, M.B., 2012. *Subprime Cities: The Political Economy of Mortgage Markets*. Wiley. Hoboken, New Jersey.

Alves, J.A., 2014. From necropolis to blackpolis: necropolitical governance and black spatial praxis in São Paulo, Brazil. *Antipode*. 46, 323–339.

Ballard, R., Dittgen, R., Harrison, P., Todes, A., 2017. Megaprojects and urban visions: Johannesburg's Corridors of Freedom and Modderfontein. *Transformation: Critical Perspectives on Southern Africa*. 95, 111–139.

Butler, T., Lees, L., 2006. Super-gentrification in Barnsbury, London: globalization and gentrifying global elites at the neighbourhood level. *Transactions of the Institute of British Geographers*. 31, 467–487.

Chakrabarty, D., 2009. *Provincializing Europe: Postcolonial Thought and Historical Difference*. Princeton University Press. Princeton, New Jersey.

Chaskin, R.J., Joseph, M.L., 2013. 'Positive' gentrification, social control and the 'Right to the City' in mixed-income communities: uses and expectations of space and place. *International Journal of Urban and Regional Research*. 37, 480–502.

Chatterjee, P., 2013. *The Politics of the Governed: Reflections on Popular Politics in Most of the World*. Columbia University Press. New York.

Claire, D., David, G., Bindi, S., 2013. Faith and suburbia: secularisation, modernity and the changing geographies of religion in London's suburbs. *Transactions of the Institute of British Geographers*. 38, 403–419.

Colomb, C., 2007. Unpacking new labour's 'urban renaissance' agenda: towards a socially sustainable reurbanization of British cities? *Planning Practice and Research*. 22, 1–24.

Comaroff, J., Comaroff, J.L., 2012. *Theory from the South: Or, How Euro-America Is Evolving toward Africa*. Paradigm Publishers. Boulder, Colorado.

Dwyer, C., Gilbert, D., Bindi, S., 2013, Faith and suburbia: secularisation, modernity and the changing geographies of religion in London's suburbs. *Transactions of the British Institute of Geographers*. 38, 403–419.

Flusty, S., 2002. The banality of interdiction: surveillance, control and the displacement of diversity. *International Journal of Urban and Regional Research*. 25, 658–664.

Fyfe, N., 2004. Zero tolerance, maximum surveillance? Deviance, difference and crime control in the late modern city, in: Lees, L. (Ed.), *The Emancipatory City? Paradoxes and Possibilities*. Sage Publications. London, pp. 40–56.

Gotz, G., Simone, A., 2003. On belonging and becoming in African cities, in: Tomlinson, R., Beauregard, R.A., Bremner, L., Mangcu, X. (Eds.), *Emerging Johannesburg: Perspectives on the Postapartheid City*. Routledge. New York, pp. 123–147.

Grosfoguel, R., 2007. The epistemic decolonial turn. *Cultural Studies*. 21, 211–223.

Gutiérrez-Rodríguez, E., 2010. *Migration, Domestic Work and Affect: A Decolonial Approach on Value and the Feminization of Labor*. Routledge. New York.

Halbert, L., Attuyer, K., 2016. Introduction: the financialisation of urban production: conditions, mediations and transformations. *Urban Studies*. 53, 1347–1361.

Hall, S.M., 2015a. Super-diverse street: a 'trans-ethnography' across migrant localities. *Ethnic and Racial Studies*. 38, 22–37.

Hall, S.M., 2015b. Migrant urbanisms: ordinary cities and everyday resistance. *Sociology*. 49, 853–869.

Harvey, D., 2012. *Rebel Cities: From the Right to the City to the Urban Revolution*. Verso Books. New York.

Houssay-Holzschuch, M., 2016. Diss and ditch? What to do with public space, in: De Backer, M., Melgaço, L., Varna, G., Menichelli, F. (Eds.), *Order and Conflict in Public Space*. Routledge. London, pp. 216–220.

Jennifer, R., 2015. 'Arriving at' urban policies: the topological spaces of urban policy mobility. *International Journal of Urban and Regional Research*. 39, 831–834.

Kihato, C., 2013. *Migrant Women of Johannesburg: Everyday Life in an In-Between City*. Palgrave Macmillan. New York.

Landau, L.B., 2018. Friendship fears and communities of convenience in Africa's urban estuaries: connection as measure of urban condition. *Urban Studies*. 55, 505–521.

Landau, L.B., Freemantle, I., 2010. Tactical cosmopolitanism and idioms of belonging: insertion and self-exclusion in Johannesburg. *Journal of Ethnicity and Migration Studies*. 36, 375–390.

Lees, L., Shin, H.B., López-Morales, E., 2016. *Planetary Gentrification*. Polity Press. Cambridge.

Lopez-Morales, E., 2011. Gentrification by ground-rent dispossession: the shadows cast by large-scale urban renewal in Santiago de Chile. *International Journal of Urban and Regional Research*. 35, 330–357.

Malcomess, B., Wilhelm-Solomon, M., 2016. Valleys of salt in the house of God: religious re-territorialisation and urban space, in: Wilhelm-Solomon, M., Nunez, L., Kankonde Bukasa, P., Malcomess, B. (Eds.), *Routes and Rites to the City: Mobility, Diversity and Religious Space in Johannesburg*. Palgrave Macmillan. London, pp. 31–60.

Maldonado-Torres, N., 2007. On the coloniality of being. *Cultural Studies*. 21, 240–270.

Mbembé, A., 2003. Necropolitics. *Public Culture*. 15, 11–40.

Mbembe, A., 2008. Aesthetics of superfluity, in: Nuttall, S., Mbembe, A. (Eds.), *Johannesburg: The Elusive Metropolis*. Duke University Press. Durham, North Carolina, pp. 37–67.

Mbembé, A., Nuttall, S., 2004. Writing the world from an African metropolis. *Public Culture*. 16, 347–372.

Mbembe, A., Nuttall, S., 2008. Introduction: Afropolis, in: Nuttall, S., Mbembe, A. (Eds.), *Johannesburg: The Elusive Metropolis*. Duke University Press. Durham, North Carolina, pp. 1–33.

McFarlane, C., 2006. Crossing borders: development, learning and the North–South divide. *Third World Quarterly*. 27, 1413–1437.

Merrifield, A., 2013. *The Politics of the Encounter: Urban Theory and Protest under Planetary Urbanization*. University of Georgia Press. Athens.

Mitchell, D., 1995. The end of public space? People's park, definitions of the public, and democracy. *Annals of the Association of American Geographers*. 85, 108–133.

Mouffe, C., 2005. Which kind of public space for a democratic habitus? in: Hillier, J., Rooksby, E. (Eds.), *Habitus: A Sense of Place*. Ashgate Publishing. Farnham, pp. 109–116.

Myers, G.A., 2011. *African Cities: Alternative Visions of Urban Theory and Practice*. Zed Books. London.

Németh, J., 2006. Conflict, exclusion, relocation: skateboarding and public space. *Journal of Urban Design*. 11, 297–318.

Nuttall, S., 2004. City forms and writing the 'now' in South Africa. *Journal of Southern African Studies*. 30, 731–748.

Parker, A., 2017. The spatial stereotype: the representation and reception of urban films in Johannesburg. *Urban Studies*. OnlineFirst.

Parnell, S., Robinson, J., 2012. (Re)theorizing cities from the Global South: looking beyond neoliberalism. *Urban Geography*. 33, 593–617.

Peck, J., 2015. Cities beyond compare? *Regional Studies*. 49, 160–182.

Pieterse, E., 2011. Grasping the unknowable: coming to grips with African urbanisms. *Social Dynamics*. 37, 5–23.

Robinson, J., 2003. Postcolonialising geography: tactics and pitfalls. *Singapore Journal of Tropical Geography*. 24, 273–289.

Robinson, J., 2006. *Ordinary Cities: Between Modernity and Development*. Routledge. London.

Robinson, J., 2011. Cities in a world of cities: the comparative gesture. *International Journal of Urban and Regional Research*. 35, 1–23.

Robinson, J. 2015. 'Arriving at' urban policies: the topological spaces of urban policy mobility. *International Journal of Urban and Regional Research*. 39, 831–834.

Roy, A., 2009. The 21st-century metropolis: new geographies of theory. *Regional Studies*. 43, 819–830.

Simone, A., 2001. Straddling the divides: remaking associational life in the informal African city. *International Journal of Urban and Regional Research*. 25, 102–117.

Simone, A., 2004. *For the City yet to Come: Changing African Life in Four Cities*. Duke University Press. Durham, North Carolina.

Simone, A., 2010. A town on its knees? Economic experimentations with postcolonial urban politics in Africa and Southeast Asia. *Theory, Culture and Society*. 27, 130–154.

Smith, N., 2002. New globalism, new urbanism: gentrification as global urban strategy. *Antipode*. 34, 427–450.

Spivak, G., 1999. *A Critique of of Postcolonial Reason*. Harvard University Press. Cambridge, Massachusetts.

Watson, V., 2014. The case for a Southern perspective in planning theory. *International Journal of E-Planning and Research*. 3, 23–37.

Winkler, T., 2013. Why won't downtown Johannesburg 'regenerate'? Reassessing Hillbrow as a case example. *Urban Forum*. 24, 309–324.

Wood, A., 2014. The politics of policy circulation: unpacking the relationship between South African and South American cities in the adoption of Bus Rapid Transit. *Antipode*. 47, 1062–1079.

Zheng, J., 2013. Creating urban images through global flows: Hong Kong real estate developers in Shanghai's urban redevelopment. *City, Culture, Society*. 4, 65–76.

Index

Note: Page numbers in *italic type* indicate figures or illustrations.